Technology & Industry
A NORDIC HERITAGE

Technology & Industry

A NORDIC HERITAGE

Jan Hult and Bengt Nyström

EDITORS

Hans Weinberger

ASSOCIATE EDITOR

Science History Publications/USA

1992

First Published in the United States of America
by Science History Publications/USA
a division of
Watson Publishing International
Post Office Box 493, Canton, MA 02021

Portions of articles included in the second part of this volume previously appeared in the journal *Teknik & Kultur*, Bengt Nystöm, editor, to whom all requests for their extended use should be addressed.

Library of Congress Cataloging-in-Publication Data
Technology & industry / Jan Hult and Bengt Nyström, editors.
p. cm.
ISBN 0-88135-151-2 (cloth) : $40.00 — ISBN 0-88135-150-4 (pbk.) : $15.00
1. Technological innovations—Economic aspects—Scandinavia-History—Case studies. 2. Scandinavia—Industries—Location—History—Case studies. I. Hult, Jan. II. Nyström, Bengt, 1937–. III. Title: Technology and industry.
HC350.Z9T47 1992
338'.064'0948—dc20 92-13610
CIP

This book has been typeset and designed by Crane Design, West Barnstable, Massachussetts, and printed and bound by Thomson-Shore, Inc., Dexter, Michigan.

Manufactured in the USA.

Contents

Acknowledgments

Publication of this volume has been made possible, in part, through the generous support and cooperation of the following:

Tekniska Museet (The Swedish National Museum of Science and Technology) in cooperation with *Uppsala University*, the *National Committee for the History of Technology*, the *Office for History of Science at Uppsala University*, the *Department of History of Science and Technology* at the *Royal Institute of Technology* in Stockholm, the *Department of Technology and Social Change* at *Linköping University*, the *Center for History of Science* at the *Royal Swedish Academy of Sciences*, all through the auspices of the *Local Arrangements Committee for the SHOT Annual Meeting 1992*

The *Central Board of National Antiquities*, Stockholm, Sweden

The journal *Teknik & Kultur*

The *National Associations of TICCIH* (The International Committee for the Conservation of the Industrial Heritage), branches in Denmark, Finland, Norway, and Sweden.

Preface

Jan Hult — Bengt Nyström

The four Nordic countries—Denmark, Finland, Norway, and Sweden—although different in character share a common history. During some periods two or three of them were joined politically. Today, they are all independent states but with close economic and cultural ties. They are modern, industrialized countries, depending to an ever-increasing degree on ties to the rest of the world.

Differences in natural resources have brought about differences in the technological and industrial development of the four countries. For example, abundant deep forests and water power in Finland, Norway, and Sweden are resources almost totally lacking in Denmark. On the other hand, highly fertile soil and modern technology have made Denmark a leading agricultural nation. Again, where Finland and Sweden have to import all their oil, Denmark and Norway are oil-producing and exporting countries.

This volume presents brief historical surveys, country by country, from which such differences and similarities may be grasped. The surveys were all written specifically for this volume and are therefore necessarily limited. They reflect their author's particular interests and are not meant to be comprehensive. Rather they should be read as attempts to sketch brief national histories of technology.

The volume also serves to make readers and visitors acquainted with recent as well as older efforts to preserve the industrial heritage of the four countries. In addition to presenting the philosophy and strategy behind these efforts, it also discusses selected technological and industrial landmarks. These monuments and sites show readers, as well as visitors, tangible industrial realities from times gone by.

Taken together, the two aspects of the Nordic technological heritage will, we hope, give the reader—and the visitor—a picture of how these countries were transformed from agrarian societies to modern industrial states in a relatively short time and without a simultaneous rupturing of their social fabrics.

We are grateful to our publisher, Science History Publications, USA, and to our associate editor, Mr. Hans Weinberger, Royal Institute of Technology in Stockholm, for helping to bring this project to its conclusion.

The Nordic region as portrayed in a map published in Paris in 1792.

Foreword

Svante Lindqvist

Anyone studying a map today tends to see *terra firma* as the natural link between towns and cities, simply because land transport is simpler and speedier than transport by sea. But the situation used to be the reverse, and in olden times, the sea was not a transport obstacle but a link. A traveller in the preindustrial era saw in a map almost a negative image of what we see today: towns and districts were linked by water but separated by land.

So although the Nordic countries may appear natural geographical units to a modern observer, this is due only to developments in transport technology in recent centuries. Previously the roles of land and sea as geographical dividing lines and connections were almost exactly reversed. It was, therefore, natural for Skåne, the southernmost part of modern Sweden, to belong for centuries to the island kingdom of Denmark. The Halland Ridge and the Småland Highland of southern Sweden were far more forbidding barriers than the Sound. For the same reason, Norway and Sweden were separated—rather than united—by the range of mountains between them, and Norway formed political and economic ties instead to Denmark and England through sea connections across the Skagerack and the North Sea. Similarly the Baltic Sea formed a link rather than an obstacle between Finland and Sweden, and the two countries were one state until the early 19th century. (To a city like Stockholm, for example, trade with the coastal towns of Finland, with the Baltic cities of North Germany, and with the ports of Estonia, Latvia, and Lithuania was a considerably simpler undertaking than trade with towns in the Swedish interior.)

As the road network and land transport improved, however, the unifying importance of water declined, and Europe's national borders began increasingly to coincide with its coastlines. Norway freed itself from its former dependence on Denmark in 1814, and Sweden lost Finland to Russia in 1809. The present political division of the Nordic landscape into the four independent nations of Finland, Norway, Denmark, and Sweden is thus fairly recent—it is, roughly speaking, about as old as macadam, the railways, and the internal combustion engine.

This may have all the appearances of a shallow metahistorical explanation, a technologically deterministic interpretation of the shifting political relations between the Nordic countries. The once superior efficiency of sea transport, by comparison with land transport, is nevertheless part of the reason why these nations—now separate—were for so long united politically, militarily, and economically in various constellations, and why they even today constitute a fairly homogeneous cultural unit.

Culturally, the Nordic countries were long part of the German-speaking sphere. Much has been written about influences from France in the 18th century, from Britain in the 19th, and from the United States in the 20th. Throughout most of their history, however, the major cultural influence on the Nordic countries has been from what is now Germany. This is also true of their technological and industrial history. Admittedly their early systems of engineering education were influenced by French schools, such as École Polytechnique in the late 18th century; their engineering industry was largely modelled on the British example in the 19th century; and since the end of the Second World War, they have been influenced by trends in the United States.

But the dominating technological and industrial influence on the Nordic countries has come from the German-speaking world. This can be seen, for example, in the Swedish mining and metallurgical technology of preindustrial times, reflected in the titles, such as *Markscheider* and *Geschworner*, of 18th century Swedish mine officials. Most major preindustrial technologies were in fact brought to Scandinavia by immigrant artisans from present-day Germany, and until the 19th century, it was primarily to the German-speaking part of the Continent that Scandinavian journeymen made their way. The strong German influence persisted into the first half of this century. In Sweden, for example, until the 1930s, the traditional Grand Tour of the final-year engineering student was almost always to Germany. Modern technologies of the interwar period, such as glass-to-metal sealing of electronic valves, were brought to the Nordic countries by artisan from the German companies Siemens and Osram. Symbolically, the earliest copy of Taylor's book on the principles of scientific management that is possessed by the library of the Royal Institute of Technology in Stockholm is not the first American edition of 1911, but a German edition of 1913, *Die Grundsätze wissenschaftlicher Betriebstührung*.

It could be argued that any general survey of the history of technology that is written anonymously (i.e., without mentioning places and names) and with no dates given more exactly than the nearest 20 years would be valid for most Western countries. However, any general history of *industrial* development must take into account geographical differences in natural resources, topography, and climate. The industrial history of the Nordic countries therefore differs from that of the rest of Western civilization largely because of geographical position of these countries on the northern periphery of Europe. Although they lie on the same latitude as Siberia, Alaska, and Greenland, the climate is nevertheless fairly similar to that of the rest of Europe, thanks to the Gulf Stream. Their severe winters have, however, always made energy technology a prime focus of technological development. Furthermore, except for the only recently exploited Norwegian and Danish deposits of oil and gas in the North Sea, the Nordic countries lack domestic deposits of fossil fuel. If a first approximation of their cultural history would be one, as argued above, that is basically similar to that of Germany, their technological history could largely be written in terms of their energy needs.

This book sets out to present these similarities and differences in technological and industrial development. A comparative international perspective is crucial to the intellectual development of the history of technology, since single-country studies risk neglecting one of the fundamental aspects of technological change: the international flow of technological knowledge and its different manifestations in various political, social, economic and cultural contexts. The embodiment of this knowledge in different artifacts, methods, processes and systems exhibits the same global multiplicity and variety as the biological species, and a comparative international perspective is essential to our understanding of the processes of evolution, mutation, and selection in the global technological habitat.

The book is being published on the occasion of the annual meeting of the Society for the History of Technology (SHOT) in Uppsala from August 16–20, 1992. It will be the first meeting of the society outside North America since it was founded in 1958, a sign of a growing awareness of the importance of a comparative international perspective. The meeting is being hosted by Tekniska Museet in Stockholm, but it is a result of the combined efforts of historians of technology and industrial archaeologists in all the Nordic countries. This book bears witness to their collaboration, and we hope that it will serve as an introduction and a stimulus to further study of the similarities and differences in our technological and industrial past.

Svante Lindqvist
Chairman Local Arrangements
Committee for the SHOT Annual
Meeting, 1992.

PART ONE

History of Technology

An Overview

Technology in Denmark

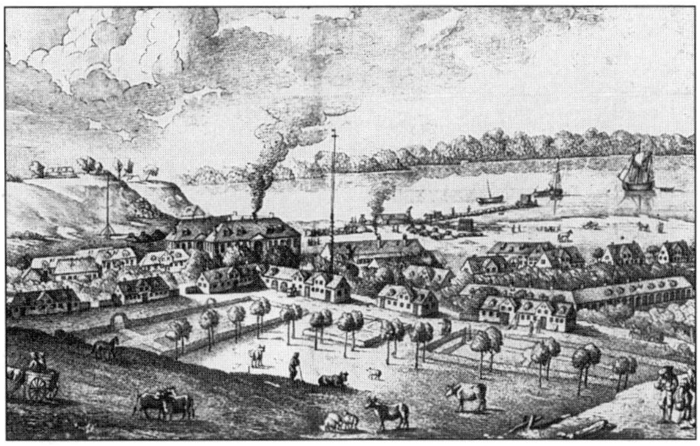

Henry Nielsen **Michael F. Wagner**

Introduction

Measured on a European scale, Denmark is a small country—and has been so for more than three centuries. Consisting of the Jutland peninsula and about 100 inhabited islands at the strategic entrance to the Baltic Sea, the land is fertile and suited for farming, but the subsoil lacks mineral deposits. And for centuries Denmark was considered short of energy resources, such as water power, coal, and oil—until the country became an oil-producing nation during the 1970s.

These facts substantiate why the principal Danish trades and industries for centuries were farming, fishing, shipping and various crafts. Denmark was a 'latecomer' in the industrialization process. Most of the technological developments in Denmark have resulted from a transfer of known and developed technologies from other countries, until 1850 mainly from England, after 1850 mainly from the USA and Germany. A common experience of most 'latecomer' nations, this in no way makes Denmark particularly interesting. *What does make Denmark unusual in a technological sense is the characteristic filtering and implementation of various technologies to render them appropriate to the particular Danish environment at any given time, locally as well as nationally.* If Denmark has contributed anything but a few ingenious inventions in specific areas to technological development, it is in the adaption, the adding of a special design and a unique style to various technological products, processes, and organizations.

1650–1814: Denmark—an Agricultural Nation

In 1660 Denmark became an absolutist state with political sovereignty vested in the king. Economic power was in the hands of the three great groups of landowners, the nobility, the church, and the Crown. Together they owned almost all land in the kingdom. The landed estate was the most important foundation for production in the country.

After a long period of prosperity, Danish society was under severe strain from ecological deterioration and economic impoverishment. Heavy mortgage and extensive sales of

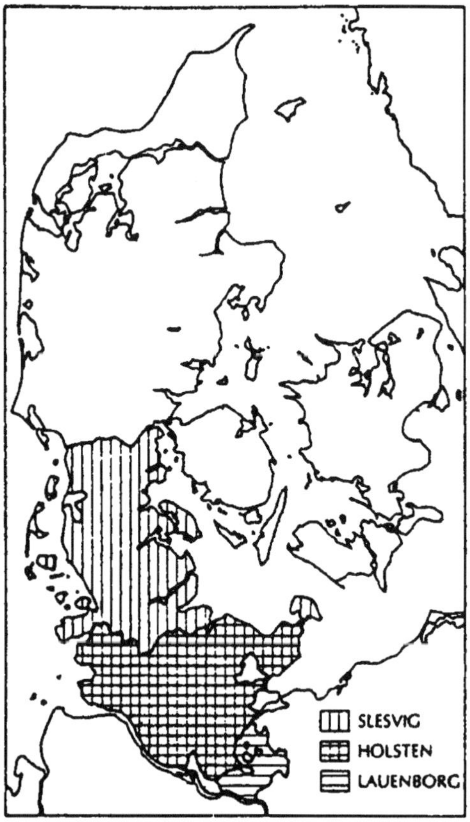

Map of Denmark with the duchies of Schleswig, Holstein, and Lauenburg. The unhatched area indicates Denmark 1864–1920. After the referendum of 1920, approximately half of Schleswig returned to Denmark.

the Crown's land took place in an attempt to make up the State deficit. The consequence of this was a profound redistribution of land and the erection of new manors on the former estates of the Crown. The result of these events was a new concentration of land and refeudalization during the first period of absolutism.

During the following century, a huge effort was made to prevent an ecological collapse. A major concern was deforestation with a shortage of firewood, fencewood, and wood for toolmaking and buildings as the result. The forests had been cut down to supply the enormous building and construction activities initiated by the Crown and the nobility during the 17th century. New forests were planted to prevent soil erosion; the ground water level was lowered by digging ditches and drainage. To improve the fertility of the soil by enhancing its nitrogen content, red clover was introduced on a massive scale. The shortage of wood forced the introduction of alternative materials; fields were fenced with stone or embanked. Tools for agriculture and handicrafts in the villages were made of iron instead of wood, and this gave the impetus for a local development of technology among the village blacksmiths.

After 1750 a period of enlightenment and agricultural reform changed the structure of the whole agricultural system. New legal and economic relations of freeholding were introduced. The old village community was gradually broken down, and the farmhouses were

TABLE 1.
The total population and its main occupations, 1800–1970. (Source: *Danmarks Historie*, Vol 9, 1990)

Year	1800	1850	1901	1930	1960	1970
Population (millions)	0.9	1.4	2.4	3.6	4.6	5.2
Record of occupations (%)						
Farming	55	44	40	30	19	10
Crafts and industry	8	21	28	29	35	34
Trade	3	4	11	11	16	16
Fishing and shipping	2	3	3			

relocated to the open farmland. The amount of acreage under cultivation expanded during the whole period. The face of the countryside gradually changed from a state of raw nature with a little cultivation in the 17th century to a cultivated landscape by 1814. Handicrafts became an important trade with decisive economic importance for the bigger towns, but at the end of the period, there were still few factories with an industrial production in Denmark.

Agriculture

According to official taxation lists from 1662, the main product in agriculture was barley. Barley requires a clay soil of high quality. Rye was common on the sandy soil, while oats were grown on the poorest soil. The peasant household was basically self-sufficient and rested on a subsistence economy. During this period a gradual changeover to a money economy took place.

The village was organized around the peasant households. The village community consisted of the common right to use the natural resources outside the village field—the common, the woods, the heath, and the moor. Also the precise coordination of work in the village field was essential. Two technological systems were employed for cultivating the land, field farming and grass farming. In field farming the village field was divided into small fields according to the number of peasants in the village, and the fields were fenced in. In grass farming the small fields in the village field were unfenced, and the cattle had to be constantly herded to prevent them from eating the crops. The crops were changed from field to field according to a preplanned system; most common was the three-field shift where crops changed from barley to rye to common in a three-year span. The crops were harvested with sickles; grass was cut with sythes.

The most important tool for the field work was a heavy wheelplough with the moldboard that had already been implemented during the 13th century. Like the sickles and sythes, this was also produced by the village blacksmith. The wheelplough was gradually improved to become a very efficient instrument, its main disadvantage being its weight. By the end of the period, four horses or oxen were still needed to pull the plough in heavy clay soil.

The innovations often came from the southern parts of the kingdom, mainly from the duchy of Holstein. After 1750 the three-field shift was abandoned on many estates, and the field were run after a Holsteinian system. This gave room for growing a number of crops for feeding purposes, such as leguminous plants, clover, and rye grass, which increase fertility and are also better fodder. To-

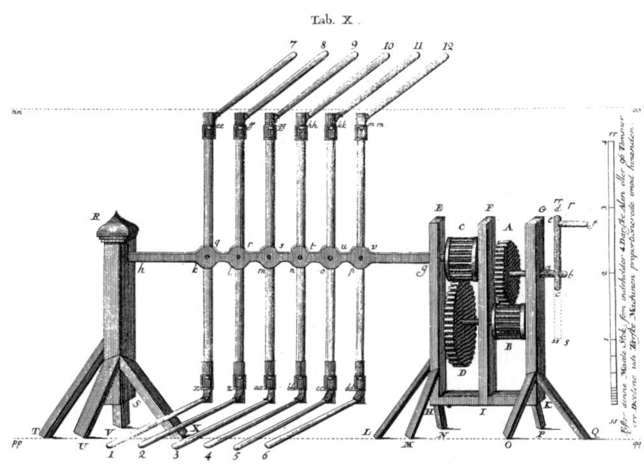

This drawing of a machine for threshing grain appeared in a publication, *Denmark and Norway's Economic Magazine*, in 1760. It was intended for the big estates, where it could save the work of 6–8 peasants and treat the straw and grain just as well.

gether with peas, wheat and potatoes were introduced as new crops for human consumption. The export of oxen was brought to a halt because of uncertain market conditions and terrible epidemics of cattle plague. Dairy farming was growing fast; it was often called 'hollanderi' because the dairy experts usually came from Holland. England and Switzerland also contributed their expertise to the development of Danish diary technology.

The reform age after 1750 resulted in substantial changes in the structure of Danish agriculture. The reforms were initiated through a process of enlightenment, in which progressive aristocrats tried to develop agricultural technology through the introduction of new ideas. Treatises on agricultural economy were published along with books on technical and productive improvements in farming. Also the legal, economic, and social position of the peasant was discussed at length. Through legislative reform, the peasant secured a new position in society. In 1769 the Royal Society for Agricultural Households was founded based on foreign models. The main purpose of the society was to "encourage the farmer, the artisan and the craftsman in all His Majesty's countries through prizes and awards."

The most important consequence of the reform age was agricultural reform, which made the peasant a free man and landowner, free of the old feudal bonds and obligations. When the farmhouses were moved from the village to the open farmland, the peasant household became socially and economically insulated from the traditional village community. By 1814 this process was well under its way.

Crafts and Industry

In the 17th and 18th centuries, most crafts were still organized according to the medieval guild system, with production methods based on the use of traditional and simple tools. In accordance with a guild's educational system, a young man began his career as an apprentice for several years in a master's shop. Having finished his apprenticeship, he was encouraged to work with other masters in Denmark as well as abroad before he could eventually hope to become a master himself. The system was rigid and patriarchal and gave little encouragement to technological innovations.

Manufacturers were already present in Denmark around 1650, and they became more numerous during the 18th century. It is important to emphasize, however, that the internal organization of their firms was never such that they would deserve the name of factories in the present meaning of the word. They were just facilities for the relatively large-scale production of certain commodities, using traditional craft techniques.

In accordance with the popular mercantilist theories of the age, the absolute monarchs and their enlightened governments of the 18th century initiated a number of manufacturing enterprises to lessen the country's dependence on foreign imports. Most of these were located near Copenhagen, the capital of the country. One of the most attractive places was at Mølleaaen, a small stream that with dams and waterwheels could deliver the necessary power to the bellows and the machinery in the new workshops. Some of the manufacturers made special textiles, such as silk; some refined sugar from the sugar cane plantations in the Danish West Indies; others produced traditional import goods, such as paper and porcelain. Iron products almost invariably came from Norway, but a few iron works did exist inside the present borders of Denmark, making guns, ammunition, and other equipment for the army and navy. By the year 1800, it was also possible to buy a few farming machines produced in Denmark, but they were only meant for the large estates of the landed aristocracy, of course.

Infrastructure

Roads. Before 1750 sea transport of people and goods was generally superior to land transport because the roads were in miserable shape. In general roads still followed old paths and trails from before the Viking Age or the Middle Ages. Having been used for oxen

Frederiksværk Gun Foundry and Powder Works. Denmark's largest industrial enterprise towards the end of the 18 century, shown in this prospect from 1773 by the painter, C.A. Lorentzen. The works were situated near a canal drawing water from lake Arresø to Roskilde Fjord on Sealand. The complex of buildings comprised a large number of worker's houses around the foundry building in the center from where the smoke rises. On top of the hill, a safe distance from the factory, are the powder magazines. A ship, probably bringing iron ore from Norway, is shown on the Fjord.

transport, they were swampy and full of holes, making them virtually impassable in heavy rains. A journey with horse cart across the country, i.e., from Copenhagen on Sealand to Kiel in Holstein, could hardly take less than a week. Many royal initiatives to improve the roads had failed.

In 1763 a new project was initiated to connect the different regions of the country by better roads. A French road engineer, J. Marmillod, was appointed to head the project, which implied the building of a new system of main roads, totalling 1,100 kilometers. From 1784 the Danish Army Engineers Corps was in charge of the construction work, which progressed slowly. The war of 1807–1814 brought construction to a half; only 250 kilometers—mainly on Sealand—had been made according to the new principles. The amount of time needed to travel from Copenhagen to Kiel via Funen and Jutland had been reduced to something like four days, but it was going to take another 50 years before the plan of 1763 was to be realized.

Optical Telegraph. To fulfill the need for fast communications among the various parts of the country, an optical telegraph line, running from Elsinore to Copenhagen, Korsør, Odense, and Fredericia to Kiel, was constructed shortly after the year 1800. A modified version of the French Chappé telegraph of 1794, it was often troubled by unstable Danish weather conditions. As a military installation, its main use was to transfer brief military and political commands between the stations on the line. The line never played a significant commercial role.

Shipping. The last half of the 18th century saw a boom in trade and shipping, partly as a result of Danish neutrality during the American War of Independence and the wars following the French Revolution. The combined tonnage of the Danish merchant fleet increased dramatically, and the owners earned huge fortunes by exploiting the transport void arising from the destruction of the belligerent powers' own merchant vessels. Danish merchant ships were allowed to transfer goods between the powers at war, sail slaves from Africa to America, and gradually take over much of the transport market for coffee, tea, cocoa, sugar, tobacco, and spices from the colonies to Europe. Copenhagen became a large European port, and Danish shipyards had a golden age. This was also true for the Danish naval yard, Holmen, in Copenhagen, which probably was the most innovative Danish 'factory' at that time. Together with other countries, Sweden and Russia in particular, Denmark followed a policy of 'armed neutrality,' which implied the escort of merchant ships by Danish warships. Thus the Danish fleet expanded at a furious pace.

The Danish government was playing a risky game with its armed neutrality policy because it was a challenge to England, the world's leading seapower. For a considerable time, everything went well, but in 1807 came the first catastrophe. Denmark suddenly found itself in alliance with Napoleon and at war with England. England retaliated immediately by exposing Copenhagen to the first massive terror bombardment in history and by conquering and/or destroying almost all the Danish warships. The immediate consequence was a period of continued expansion for the infant Danish industry because the war almost completely halted foreign imports and because the army and the navy placed large orders. Napoleon's defeat in 1814 brought another catastrophe, partly due to the loss of Norway and partly to the great problems in finding export markets during the long European depression following the Napoleonic Wars.

1814–1920: The Century of Change

After the surrender of Norway, Denmark experienced a severe economic crisis until a period of moderate economic growth set in after 1829, when England abandoned the corn laws and thus resumed the import of Danish grain to the English market. The domestic need for grain also increased because of the urbanization taking place in the 19th century, especially during its second half. The towns grew tremendously. The century-old city limits of Copenhagen fell in 1871, and this event marked the beginning of the capital's sprawling growth, which has continued ever since. In 1896 Copenhagen had over 400,000 inhabitants, more than half of these being newcomers from the provinces.

During the 1880s and the 1890s, Danish agriculture was forced into a difficult changeover from grain production to animal production. This took place at the same time as industry began to grow rapidly, eventually making urbanization and the development of town-based occupations the most important motive power behind societal changes. During the whole period, farming preserved its position as the country's main trade, employing the greatest share of the population and contributing most of its GNP.

Politically the period was characterized by the abolition of the absolute monarchy in 1848, and the first Danish constitution in June 1849, marking the introduction of democracy.

1857 was an important year for Danish trade and industry. That year legislation abolished all monopolies, which for centuries had been associated with the guild system, and which had been a very efficient barrier against free trade. Second, Øresundstolden ('the Sound Customs'), which added considerable cost to all exported and imported goods, were removed. And third, Privatbanken ('the Private

Bank') was founded to procure risk capital for the development of private industries. These provisions were important for the industrialization process that began in the 1860s and gathered momentum in the 1880s and 1890s.

Following a period of bourgeois rule for one and a half decades towards the end of the century, Venstre ('the Left'), the liberal farmer's party, came to power in 1901 with support from the rapidly growing worker's movement. The Social Democratic worker's movement, founded in 1871, campaigned successfully for the organization of skilled craftsmen and unskilled industrial workers in trade unions. As a countermeasure, the employers formed a central organization in 1896, and in 1899 the two parties negotiated a compromise concerning future collective bargainings. This settlement recognized the right of employers to manage and share the work. The state participated as a third party in the labor market with the role of arbitrator, which marks the beginning of the economic organization later to be known as 'the Scandinavian model.'

In the 19th century, Denmark was twice at war with Prussia and Austria, in 1848–1850 and again in 1863–1864. The last war clearly revealed Denmark's weakness in a changing Europe. The result was total defeat, the duchies of Holstein and Schleswig were lost to the rapidly expanding military and economic power in Central Europe, Germany. In the First World War, Denmark remained neutral. After the war, as a result of the Versailles Treaty, a referendum took place in Schleswig in 1920, and half of it (southern Jutland) returned to Denmark.

Agriculture

The agricultural reforms laid the foundation for a class of independent peasants (freeholders). This was to have a major influence on Danish culture and society in the 19th century. The new class of farmers displayed great interest in technological development to improve their agricultural production. It was a primary force in the history of Danish technology and society.

In 1814 grain was still the most important farm product, especially barley, but wheat became very popular during the 'grain sale period' from the 1830s to 1875. Many experiments with foreign wheat were made, not all of them equally successful. New technology was introduced on a growing number of farms. The modern American swing plough increasingly replaced the old wheelplough. However, conservative farmers cherished the old plough, and improvements were still being made on it well into the 1860s.

The demand for better and more varied and specialized tools for tilling fostered the industrial production of agricultural tools. Mechanic O.J. Winstrup established a factory in Copenhagen shortly after 1814 and was able to compete with imported tools. From 1836 agricultural tools were also made in Odense at the M.P. Allerups iron foundry; in Aarhus Søren Frich started an iron foundry and engine works that was to deliver 50 percent of its production to farmers in eastern Jutland. In most Danish towns, industries were based on iron foundries after 1850.

Around 1860 half the population were peasants. Although there had been an enormous growth in population during the 18th century and the first half of the 19th century (see Table 1), it did not considerably change the balance between countryside and town. The agricultural sector was able to employ the many new hands by putting more land under the plough and by increasing production. The increase in farming also increased the number of farmhouses with little or no land attached to them. This meant the creation in the countryside of a class of farm laborers without land. It was from this rural working class that the rising industries in the towns recruited the labor force that was needed for the new factories.

As an offspring of the reform age, a commission of education had been founded in 1789. Its main task was to reform the elementary schools in the countryside. With the general reform law for the schools in 1814, elementary school was made compulsory for all children between ages 7 and 14. Even though the curriculum largely consisted of beatings and reading the catechism, the foundation for national literacy was prepared by this legislation.

The Danish peasant class were to become the cultural determinant of Danish society during the 19th century. An important factor in this process was the folk high schools that started after 1844 in many of the provincial towns. The folk high school was aimed at adult education, but it was not a practical school of agriculture nor a theoretical Latin grammar school. The young sons and daughters of the farmers frequented these schools and were taught a blend of nationalist ideology, Nordic mythology and history, Christianity, and modern rationalism. The priest and poet N.F.S. Grundtvig was the spiritual founder of the folk high school movement, which was to become an important factor in the evolution of the farmers' cooperative movement in the 1880s.

In 1858 a scientific school for agriculture, Den kongelige Veterinær og Landbohøjskole (the Royal Danish Veterinary and Agricultural University), opened at the University of Copenhagen. This was the institutional basis for the development of science-based agricultural technology in Denmark. At the same time, a number of farming schools were also opened to educate young farmers through short courses in scientific and practical agriculture.

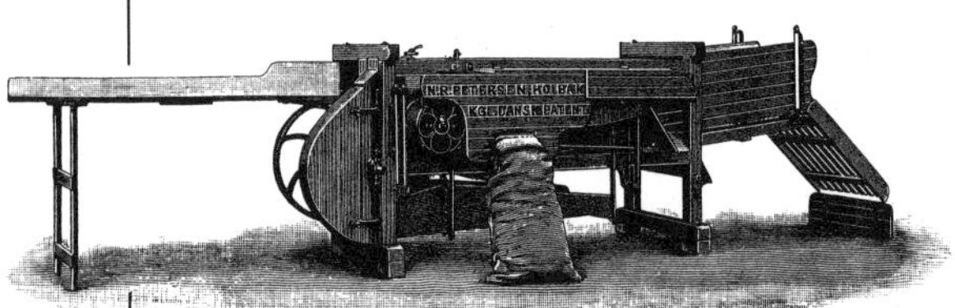

Just before 1900, a Danish machine factory, N.R. Petersen in Holbæk, succeeded in developing a horse-driven, self-cleaning threshing machine able to clean the corn as well as the larger and more much expensive English steam threshing machines could. The Danish-produced machine was bought by many farms.

After 1860 the bigger farms began to increase their numbers of cattle and switched over to animal products instead of grain for export. This development was beneficial because it increased the amount of manure and improved farm productivity. A rise in the price of manufactured animal products and a drop in the price of grain on the world market helped to accelerate this process.

In Danish agriculture, 1860–1920 is characterized by technological change. The switch from grain to animal products is obvious, but behind it lies a profound technological change in all of Danish agriculture. Traditional Danish breeds of animals and varieties of grain were commonly grown in 1860. By 1920 Danish farmers planted seeds and raised animals that had been scientifically refined and improved. The total acreage of farmland was increased through massive projects of land reclamation. Most spectacular was the cultivation of the heath in Jutland after 1866 and the drainage of the fjords in the 1870s. From 1861–1881, the cultivated farmland increased about 15 percent. A new infrastructure was created with the building of the railways in Jutland. The railway made it possible to transport fertilizer to the farmers and export farm produce to the world market.

The Danish farmers' connection with the world market was improved in several ways. From the U.S. there was a growing transfer of technology to Danish farmers, which gave them the opportunity to produce Danish copies of the same technology. This is characteristic of the development of the swing plough in Denmark. Around 1860, although the swing plough was commonly used on the farms, most of these ploughs were made by local blacksmiths. In 1886 the first factories started to manufacture swing ploughs on a massive scale. This was followed by the systematic development of Danish models. At the yearly plowing contests, a Danish model was first rated better than the imported American models in 1904. Harrows and rollers were also imported, improved, and sold as 'Danish models.' From 1890 there was considerable production of sowing machines. The tools for tilling the soil were also improved after the American pattern, which meant the general introduction of lighter and stronger materials in the production of hand tools.

Mechanization of farm work became widespread. After 1890 it was common practice to hire a traction engine and a threshing-machine on the smaller farms. In return for the mechanization of the heavy field work came new work-intensive and time-consuming jobs; by 1914 a lot of manpower went into rinsing beets and milking cows. Several experiments with machines for milking cows were conducted, but without any satisfying results in this period.

The workforce in the countryside only increased by 20 percent after 1860. At the same time, animal and mechanical power on Danish farms grew substantially. The total number of horses increased by almost 60 percent, and every horse spent many more hours working in the new machines. Mechanical power increased after 1890, when steam engines, explosion engines, windmills, and dynamos were introduced on many farms.

A whole new agricultural trade, wood plantations, was introduced around 1866, when the Danish Heath Society (Hedeselskabet) was started with the declared purpose of reforesting the poorer soil, which was unfit for other farming. The leader of the Hedeselskabet, Enrico Dalgas, epitomized the productivist spirit that sprang out of the military defeat by Prussia and the severe loss of territory that followed the war of 1864 with the words: "What is outwardly lost, must be inwardly gained." The Hedeselskabet embarked on a project of intensive planting and forestation that finally repaired the damages made over several centuries of malpractice in the forests. Through this effort, the acreage of Danish woodlands was almost doubled by 1920. In many ways this is also the epitaph for the general development of Danish agriculture from the late 1860s. Much new land was reclaimed and put under the plough after the loss of Schleswig-Holstein. But in the long run, the most important agent for the development of an agricultural technology in Denmark was perhaps the farmers' cooperative movement that evolved after the War of 1864.

The Cooperative Dairy. The development of cooperative dairies is a fine illustration of the way Danish scientific institutions and traditional handicrafts could work together to develop new technology. The first cooperative dairy, *Andelsmejeri*, was founded in Hjedding in western Jutland in 1882. Hjedding Andelsmejeri was soon to be followed by many others. In 1887–1888 alone, more than 400 cooperative dairies mushroomed all over the country. By 1890 Denmark was covered by cooperative dairies, and around 30 percent of the milk producers were part of the movement and delivered all their milk to the cooperatives. At this time many of the big farms gave up their own dairy and joined the cooperative dairies.

Traditionally in the peasant household, the dairy was worked by women. Most farms had only one to three cows, so the amount of milk for dairy produce was small, and the work not very time consuming. But the production of butter took a long time, and the resulting 'peasant-butter' was often harsh tasting and ill-reputed. When the number of cattle began to increase during the 19th century, the demand for better quality butter rose accordingly. This meant more work in the dairies, and some women specialized in the manufacture of dairy produce. Thus the production of butter and cheese had evolved into a woman's trade with a corresponding high degree of economic independence.

In the 1830s many of the big estates had invested in the development of dairies. The Society for Agricultural Economy opened a school in 1836 with a two-year education for dairy women. The new dairy women went to work in the new dairies of the big estates. The result of their efforts was better dairy products. Some of the women even wrote books about the best ways to make cheese and butter. Peasant-butter still had a bad reputation and was sold for the lowest price. 'Estate-butter' got the highest price. Its production time was much shorter due to the bigger amount of milk from the estate. And the big farmers invested in new buildings and the most modern methods of dairy production.

After 1829 there was a remarkable improvement in dairy products from estates in Holstein. These estates had switched to a new method where the cream was skimmed after the milk had been placed in open, flat tubs. This method was soon copied by other estates in the rest of the country. The result of this improvement was an increase in the production of estate-butter. After the war of 1864, scientific analysis of dairy farming was conducted by the chemical laboratories of the Landbohøjskolen, where scientists developed new methods and techniques for making dairy products during the 1860s.

The technological breakthrough to modern dairy farming came in 1879 when L.C. Nielsen constructed the Maglekilde-centrifuge. Nielsen's centrifuge works continuously, and a series of experiments at the Landbohøjskolen had proved its superiority to other methods for skimming cream. The amount of butter per kg. of milk was much higher, but even more important, was the fact that the centrifuge made it possible to aggregate the milk from different farms without any loss in output, which was impossible with any other known method. The technological conditions for dairy production on a common or cooperative level were thus at hand, and this opportunity was seized by the Danish peasants. Burmeister & Wain obtained the rights to manufacture the Maglekilde-centri-

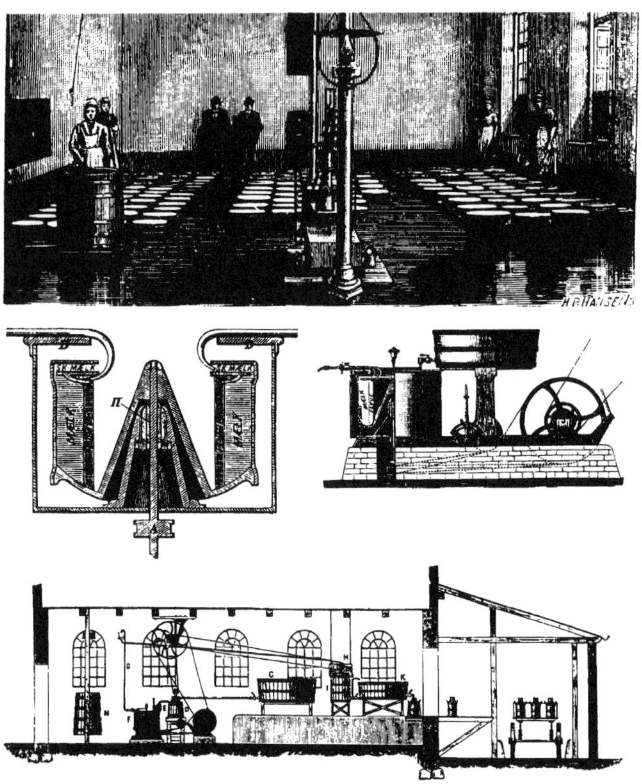

The technological development of the Danish dairy industry in the 19th century is shown here. On top is a Holstein dairy around 1860 in a specially designed building, the flat tubs on the floor and the dairy women performing their task. In the middle L.C. Nielsens centrifuge from 1879 is shown. The bottom picture shows the complete machinery of a typical cooperative dairy around 1885.

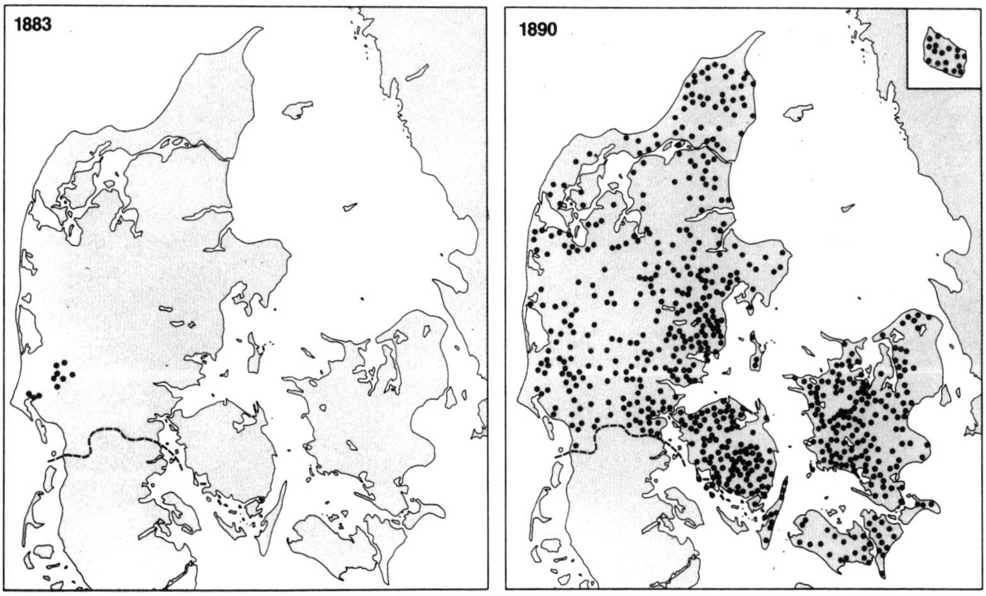

Map showing the dissemination of cooperative dairies in Denmark, 1883 and 1890.

fuge in 1881. The Swedish Laval separator was developed at approximately the same time and these two types of machines quickly spread among Danish dairies. In 1881 there were 150 centrifuges working in Danish dairies; by the end of 1882, there were between 500 and 600. By the beginning of 1885, a reliable estimate talks about more than 1,300 centrifuges being used in Danish dairies.

There existed a massive danger of infection from cattle tuberculosis from production based on aggregated milk from many different farms. The skim milk was returned to the farmers for feeding calves. Cattle tuberculosis was a serious threat to man and beast alike, and through these channels of distribution it could spread rapidly. Through intensive research at the Landbohøjskolen, the problem was solved by pasteurizing the skim milk at 80 degrees Celsius. The aggregate milk also ran the risk of going sour, but pasteurization of the cream solved this problem. Inspired by the research laboratories at Carlsberg, that had made culture yeast in the early 1880s, the laboratory at the Landbohøjskolen succeeded in isolating and growing lactic acid. After 1890 it became possible to implement these results in dairy production. By 1898 a law was enacted to control cattle tuberculosis. It became compulsory to pasteurize the skim milk as well as the cream for producing butter.

The Danish dairy industry embarked on a process of scientification and technologization during the 1880s. Dairy work changed from being a woman's trade to a man's trade. This change did not come without conflicts between the sexes. The dairy women had accumulated considerable knowledge of dairy tradition. But the cooperative dairy movement simply refused, with few exceptions, to employ women in their dairies; they would rather employ an unskilled man than a woman expert with long practical experience behind her. The explanation given for this was that women did not go well with machines and had no understanding of scientific methods.

Taken as a whole, the cooperative peasant movement was based on a different principle of organization of ownership and management, but not on an alternative technology, from that employed by a capitalist-owned and managed industry. But in the case of the cooperative dairies, the picture is different. There, a completely new technology, the centrifuge, allowed a major breakthrough to occur for Danish dairy production with a high and standardized quality as the result. But this technology did not mean that the organization of the dairy had to be cooperative in nature. This specific outcome is a result of how Danish society at a general level was developing at this period. The Maglekilde-centrifuge, the laboratory at the Landbohøjskolen, the cooperative peasant movement, the Danish economy, and the world market were spun together by a strong thread that still binds essential parts of the Danish economy together today.

Danish agriculture found a short cut from scientific development to practical farming. Several things helped this process. The general level of education among the peasants was relatively high with a strong enthusiasm for education. The many new folk high schools and agricultural schools made it possible to fill this need. Peasants organized in professional societies, hired consultants, held meetings and cattle shows, gave prizes for the best products, and worked to improve production together with the Landbohøjskolen in Copenhagen. That all these combined efforts were successful is evident from the official statistics for milk production. In the 30 years from 1861 to 1891, the average Danish cow doubled her yield from 1,000 to 2,000 liters of milk a year.

Crafts and industry

The loss of Norway in 1814 left Denmark almost without iron works—which thus had to be established before any large-scale industrialization could take place in Denmark. The economic depression from 1814 to 1830 accentuated the problem.

Around 1840 Danish industry was still small and weak. Labelling businesses with fewer than six employees as crafts (which is common practice in Denmark) and with more than six employees as industries, the number of industrial firms in 1839 in or near Copenhagen was only a few hundred, the total number of employees being only 6,600 as against nearly 40,000 craftsmen. Most industries were located in the Copenhagen area and were organized as manufacturers, textile manufacturers being the most typical. Other

TABLE 2.
The total labor force and mechanical power in Copenhagen industrial works, 1839–1914. (Source: Hyldtoft, 1984).

	Work force in industry		Mechanical power in industry		
Year	*Number*	*Growth per year*	*Number of factories*	*Power (hp)*	*Growth per year*
1839	6,600		14	183	
1847	8,500	3.2%	47	386	9.8%
1855	10,600	2.8%	82	819	9.9%
1873	21,400	4.0%	182	2,380	6.1%
1897	44,200	3.1%	780	11,500	6.8%
1906	52,000	1.8%	1260	25,400	9.2%
1914	61,500	2.1%	2100	51,100	9.1%

important works were the naval shipyard, Holmen, the iron foundry and gun factory at Frederiksværk in north Sealand, the Royal Porcelain Factory in Copenhagen, and a number of paper mills. The level of mechanization was very modest, power being mainly provided by animals, waterwheels, and windmills. In 1820 only four stationary steam engines were installed in Denmark, with powers ranging from 8 to 20 hp. All four engines had been imported from England. A survey of 1839 by the new Danish Industrial Foundation, showed the number of stationary steam engines in all of Denmark to be only 22. Half of them had been built in England, the rest in Denmark.

One reason why the craft industry remained the most important occupation (with the exception of agriculture) must be sought in the network of technical schools emerging in all Danish towns after 1870. These schools did not materialize as a result of any initiative from 'the top.' Most of them were founded by craft organizations or private persons. At the turn of the century, more than 100 technical schools existed, with over 20,000 students. The Apprentice Law of 1899 recognized the technical schools as important for the education of every apprentice.

Technical education at the highest level was offered by the Polyteknisk Læreanstalt, a technical university founded in 1829 by the Danish physicist H.C. Ørsted, who also became its first director. The teaching at Polyteknisk was rather theoretical, and it did not have much immediate impact on technological development in Denmark. From 1829 to 1890, about 500 people graduated from Polyteknisk, most finding jobs abroad or in the public sector, which expanded considerably after 1850. During the next ten years, from 1890 to 1900, another 500 persons graduated. It was only at the beginning of our century, however, that the role of this school in the technological development of modern Danish society became more visible, as engineers from Polyteknisk took over more and more innovative jobs in private industry.

But there were other kinds of engineers. Building on the practical experience of their students, the best technical schools offered short, specialized, and practical engineering courses, thus creating the so-called Teknika ('Technical Engineering Colleges'), which have also had a significant impact on Danish industry in the 20th century.

It is impossible, of course, to give a detailed account of the multifaceted technological development inside the Danish industrial sector between 1814 and 1920. Consequently we have chosen to provide short narratives of one person and two companies as examples of the driving forces behind essential inventions and technological innovations: Søren Hjorth (inventor), Carlsberg (brewery), and B&W (shipyard and motor factory).

Søren Hjorth had no formal technical training, but his great interest in technology led him to attend a number of H.C. Ørsted's lectures at the Polyteknisk in the late 1830s. In 1842 he conceived the idea of an electric motor that could replace the well-known steam engine. During a stay in England in 1849, Søren Hjorth took out an English patent for his invention, and he succeeded in raising economic and technical assistance to realize his invention. His engine worked so well that it won first prize at the World Exhibition at the Crystal Palace in London in 1851.

The current for Hjorth's electric motor was provided by the usual galvanic elements, so-called 'wet batteries,' but many batteries were needed because they had to be constantly replaced. Hjorth's motor could not compete with the steam engine until a better source of current became available. Thus it was only natural to look at the magneto-electric machines (dynamos) available at that time. Hjorth did that—and soon found that they were not suitable for the job, partly because they used permanent magnets that gradually lost their strength.

However, Søren Hjorth thought he had a solution to this problem, too. His idea was to build a 'dry battery,' a dynamo that besides permanent magnets was also provided with cast-iron electro-magnets that were activated by the current generated by the machine itself. The idea was first expressed in his notebook in 1851 and was gradually developed over the following years. As the prospect of having his invention developed in Denmark, an industrially backward country, seemed too small, he returned to England. In 1855 he was granted a patent and persuaded a company, Malcolm & Campbell in Liverpool, to build a prototype of his machine. Thus, at the end of 1855, he had at his disposal a prize-winning electro-motor as well as a dynamo constructed according to principles more advanced than any other dynamo at that time.

However, what looked like another peak in Hjorth's rising career turned out to be the beginning of a fast and sad decline. The dynamo did not function properly; the money from Denmark was soon spent; and all attempts to raise new risk capital were in vain. At this critical moment, no one believed sufficiently in his invention to provide him with the financial support for a continued and perhaps expensive development program. As with so many other inventors before and after him, Hjorth had to learn that good ideas do not guarantee success. In 1856 he gave up and returned to Denmark a poor and disillusioned man.

Instead of Hjorth, the German engineer and entrepreneur Werner Siemens (together with two English inventors, C. Wheatsone and S.A. Varley) in 1866 was recognized as the inventor of 'the dynamical principle.' Siemens had an enormous advantage over Søren Hjorth. His company already produced dynamos, and he had no need to beg investors for financial support. His own well-consolidated company, Siemens & Halske, was able to develop and market the new dynamo quickly and efficiently.

Carlsberg. In 1835 more than 100 breweries existed in Copenhagen. Until 1805 they had been linked together in a brewer's guild in which the different breweries did the brewing by turns, a system that effectively prevented any product development. Although the guild had been abolished in 1805, this step had not caused any substantial changes. The normal Danish beer continued to be top-fermented beer, and the quality was rather poor. However, after the young Danish brewer J.C. Jacobsen in 1836 had an opportunity to taste bottom-fermented German lager beer, he undertook several journeys to Bavarian breweries to study the fabrication process—and to get hold of the necessary yeast. Following a series of experiments with lengthy bottom fermentation at a low temperature, he applied for—and got—permission to build a new brewery in Valby, just outside the old walls of Copenhagen.

Jacobsen's new beer became popular, but still did not always taste as it should. The problem was well known among brewers and was associated with the yeast. The growing production and the growing competition had forced the breweries to brew beer the year round, which had not always been the case before. Summer brewing was particularly difficult because germs and wild yeast reproduced rapidly during hot summer months, giving the beer a bad smell and taste.

To solve these problems, Jacobsen in 1857 founded a research laboratory at the Carlsberg Brewery in Valby. From the start, the laboratory was divided into chemical and physiological departments. The head of the chemical department, Johannes Keldahl, in 1883 developed a method to determine the nitrogen content in organic compounds. Of even greater commercial importance were the lengthy research and development carried out by Emil Christian Hansen, head of the physiological laboratory. This culminated in 1883 in the construction of a yeast-growing apparatus in which the yeast for beer production was allowed to develop from one cell of a proper yeast sort ('culture yeast'). Hansen's method has since become the standard yeast-growing method at all breweries throughout the world.

Burmeister & Wain (B&W). In 1898, B&W built the first 20 hp diesel motor in accordance with blueprints from Rudolf Diesels' German company 'Algemeine Gesellschaft für Dieselmotoren.' In the following years, many experiments were carried out with this machine, particularly to try out various fuel-injection techniques. B&W's diesel engine worked properly, but nevertheless the company's management were against producing diesel engines because fuel (kerosene) was estimated to be too expensive to make the motor competitive in Denmark. This view was changed when heavy oil products became available and when the B&W engineers succeeded in adapting the motor to run on heavy, but cheap, imported oil. B&W chose to build a new diesel engine that was a cross between a German and a Swedish model. Although also aimed at private industry, it was above all produced to satisfy a growing need for a suitable power engine at the many small electric power stations springing up like mushrooms all over the country during the first two decades of the 20th century. The first B&W diesel engines rated from 8 to 40 hp, but already in 1909, the company was able to deliver four 150 hp diesel engines to a new electric power company in northern Jutland.

In 1909 Diesels' general patent expired, and licencees all over the world were thrown into fierce competition over designs and uses of diesel engines. B&W made the decisive gamble that made the company world famous, and turned it into a leading maker of large diesel engines for years to come. B&W took the risk of building a 7,400-ton diesel motor ship, more than ten times larger than any diesel ship ever built before. Ordered by the Danish East Asiatic Company, the ship had two four-stroke, eight-cylinder engines of 1,250 hp each. The name of the ship was *Selandia.* No radical new ideas were involved in its construction, but it was a large-scale work in which the most modern German, Swedish, and American technology was transferred to B&W and welded together in a carefully prepared and well-functioning union. The ship's maiden voyage caused a storm of interest in maritime circles all over the world, and ensured B&W full order books for many years.

Infrastructure

Railways. While the Danish network of new roads was slowly being built during the 19th century, the railway system progressed at a much quicker pace. Construction started in the 1840s, and all the main lines were finished by 1875. The actual siting of the railway lines and the sequence in which the construction took place were a textbook example of how national and military considerations can overrule technical and economical arguments.

The first Danish railway was established in Schleswig-Holstein between Altona (next to

The diesel motor ship *Selandia* returning to Copenhagen from its maiden voyage to the Far East in 1912. To most spectators the most astonishing thing about *Selandia* was its lack of smokestacks!

Hamburg on the North Sea) and Schleswig (on the Baltic). The main reason for this seemingly strange priority in a remote part of Denmark was that the Danish government wanted to anticipate a German railway planned between Hamburg and Kiel because it feared the growing German influence in Schleswig-Holstein. Thus the Altona-Schleswig line was heavily subsidized by the Danish state and was inaugurated in 1844. That same year the construction of a railway was begun on Sealand, but when the initiators (Søren Hjorth was one of them) applied for state subsidies, they were turned down. Consequently the Sealand Railways became a private company. Much of the capital was raised in Germany, but the contractors were English, and the technology was mostly English, too.

When the Jutland railways were about to be established, Denmark had just got a new, democratically elected government, and the anti-German feelings after the war of 1848–1850 were at their climax. One of the consequences was a long and bitter argument between different groups in favor of one or another of several railway proposals, the most radical being to build a number of cross railway lines from the east coast to the west coast, but no lines from north to south. The idea was that Denmark should orient itself towards England, but turn its back demonstratively to Germany!

The final decision was passed by the Danish parliament in 1861. The legislation gave the English entrepreneurs Peto, Brassey & Betts the concession to build and run a north-south main line running from town to town in eastern Jutland and in addition two cross lines from east to west. The work proceeded according to schedule, but when the entrepreneurs found themselves in serious economic troubles during the construction, parliament decided that the Danish state should take over financing and running the main lines in Jutland. Private interests were still allowed to run local railway lines, however. In 1880 the state took over the railways on Sealand and thus because the owner of more than 80 percent of all Danish railways. From 1883 a connection between the railways on Sealand and those on Funen and Jutland was established through a state-owned railway steam ferry across the Great Belt.

The increasing amount of goods to be transported between east and west Denmark and between Denmark and other countries was mainly done by sea. Many existing ports were enlarged, and on the west coast of Jutland, at Esbjerg, a new port was founded. Construction took place between 1868 and 1878, and this new port quickly became the largest export harbor for Danish farm products. The ships also went through a technological development: the first steam ships to be built in Denmark date back to the 1830s, and in the 1840s, the first iron ships appeared. As late as in 1913, sailing ships still made up 20 percent of the total tonnage, but about the same time, diesel motor ships appeared as serious competitors to the steam ships.

Telegraphy and Telephony. G. Schram, director of the newly finished railway from Copenhagen to Roskilde, in 1840 proposed the construction of an electromagnetic telegraph line between the two cities. He further proposed its continuation to the Great Belt, the island of Funen, and Jutland. The Danish government was not enthusiastic about new money-consuming proposals, but the military strongly recommended it as a substitution for the unreliable optical telegraph. Since the English authorities also pressured the Danish government to establish a telegraph line from Elsinore and Copenhagen to Hamburg (a connection between Hamburg and London already existed), the line was finally constructed in 1852–1854. In charge of its construction was Captain Lehmann of the Danish Army Engineer Corps. The material was delivered by relatively unknown English and German companies, and this fact in connection with the inexperience of the Danish engineers led to many problems in the beginning. After a number of years, much of the underground cables were removed and replaced by air telegraph wires. The daily operations were led by a state-owned company, the Danish State Telegraph that gradually developed considerable competence and skill in its field, not least through an extensive staff education program. After 1870 the state telegraph had almost a monopoly on telegraphic traffic in Denmark.

Thus Denmark was a latecomer to the most sophisticated technology of the time, the electromagnetic telegraph technology. But despite this, a Danish company, Store Nordiske Telegrafselskab ('Great Nordic') became a global heavy weight in the telegraph business. The company was founded in 1869 as a fusion of three Danish-English-Norwegian companies owning seacables between the three countries in addition to a cable between Copenhagen and St. Petersburg in Russia. The initiator was C.F. Tietgen, director of Privatbanken, which was the largest Danish bank at that time. Tietgen's vision with Great Nordic was to turn Denmark into a telegraphic crossroads between England and the USA in the west, Russia in the east, Scandinavia in the north, and Central and Southern Europe in the south. Based on a thorough analysis of the geopolitical situation around 1870, Tietgen's plan assumed that many countries—England foremost—were unhappy with the prospects of having a united, aggressive Germany as the center of European telegraph communications. Tietgen reasoned that a Danish company that could offer safe, reliable, and competitive transit traffic of telegrams through Danish territories and via Danish-controlled seacables would find many customers. Tietgen was proved right when Great Nordic in 1870–1872 became the first company in the world to establish telegraphic connections between Europe and the Far East. Besides the European cables, Great Nordic laid cables between Vladivostok in Russia, Nagasaki in Japan, and Shanghai in China. And Tietgen's achievement was crowned when he—after intense lobbying in St. Petersburg—succeeded in obtaining a remarkable agreement with the Russian government, giving Great Nordic the concession to transfer telegrams from the Far East to Western Europe via the just finished Russian telegraph line from St. Petersburg across Siberia to Vladivostok.

The first telephone exchanges in Copenhagen were set up in 1879. Two private companies were competitors on the market, but already in 1882, they were fused into the Copenhagen Telephone Company Ltd. under C.F. Tietgen's leadership. The telephone companies in Copenhagen as well as in the provinces were soon nationalized, however. A new Concession Law of 1894 guaranteed the state a controlling interest in all Danish telephone companies.

Wireless Communication. In September 1902, Valdemar Poulsen, Danish telephone technician, applied for a patent on a new method to generate continuous electromagnetic waves in the khz and Mhz frequency regions. During the following 10 to 20 years, Poulsen's method revolutionized wireless communication, until then almost identical with the spark technology associated with Marconi. The new Poulsen technology, known as 'the electric arc' or 'the Poulsen Arc,' passed through its first intensive development in Denmark in 1902–1908. At the center of the development were the practically oriented inventor, Valdemar Poulsen, and his collaborator, the more theoretically minded engineer, P.O. Pedersen. In 1908 Denmark probably possessed the most advanced experimental communication equipment in the world: an efficient Poulsen Arc consuming a few kW of power, and a detector of radio waves—called 'the tikker'—that was much superior to the usual Branly coherer. Using this equipment, Poulsen and Pedersen were able to maintain a reliable wireless telegraphic communication between test stations in Copenhagen, Esbjerg on the west coast of Jutland (Distance 300 km), and Newcastle in England (distance 900 km). Tests with wireless telephony, impossible with spark technology, were initiated in 1906.

The further development of the Poulsen technology did not take place in Denmark,

Poulsen station for wireless telegraphy and telephony in Lyngby near Copenhagen, 1909. On the table (left) is the Poulsen Arc, then the switch activating the transmitter/receiver; just to the right of the table's midpoint are the microphones and furthest to the right the telegraph key.

however. It was the German company, C. Lorentz (which had acquired the patent rights for Germany), and the American subsidiary company, 'the Poulsen Wireless Telephone and Telegraph Company,' led by the engineer C.F. Elwell, that developed the Poulsen system into gigantic Poulsen stations consuming hundreds of kW. Perhaps the most important reason why the initiative went to the Germans and the Americans is that while the Danish navy was reluctant to invest in the new communication system before it was a proven technology the German and American navies did not have similar reservations, but gambled—and won.

Electrification. The electrification of Denmark came late. Many towns had established municipal gas stations supplying gas for street lights as well as various domestic services, and now the authorities feared that an investment in municipal lighting stations would only mean serious competition for their own gas stations. Thus nothing happened, and the press wrote sarcastic articles about the usual Danish backwardness. Several private companies, for example B&W and Carlsberg, lost their patience and established their own lighting systems.

Private persons were in fact also behind the first Danish central stations. On October 21, 1891, a small private DC station in Køge south of Copenhagen, started servicing its customers (totalling only 100 incandescent lamps) with 100 V current. During the following months and years, 'lighting stations' sprang up in Copenhagen and all the major towns. Most of the stations were privately owned, but some were owned by municipalities. In 1904 three stations with a total power of 2 MW existed in what is today the center of Copenhagen, all of them municipal owned. Half of the power was used for incandescent as well as arc lights; the rest was used to run the electric street cars that were introduced in Copenhagen in 1902. Since these municipal stations were soon proved to be profitable businesses and since their growth did not imply declining gas sales, most local authorities took over the private electric companies as soon as their concessions expired.

The early phase of Denmark's electrification had—like almost everywhere—a touch of anarchy. Large and small, public and private power companies occurred side by side, as did different primary movers, such as gas motors, steam engines, diesel motors (from 1905), and different supply voltages, different price policies, etc. Such were the conditions until about 1905 when the first centralizing tendencies began to appear in eastern Denmark, partly due to pressure from the first generation of professionally trained electric engineers from abroad and from the Polyteknisk in Copenhagen (specializing in electric engineering became possible in 1903). The joint public and private company in North Sealand, NESA, initiated the centralizing process by inaugurating Denmark's first station for high-voltage alternating current in Skovshoved. The primary voltage was 10 kV, which was transformed into a secondary voltage of 220 V before reaching the customers.

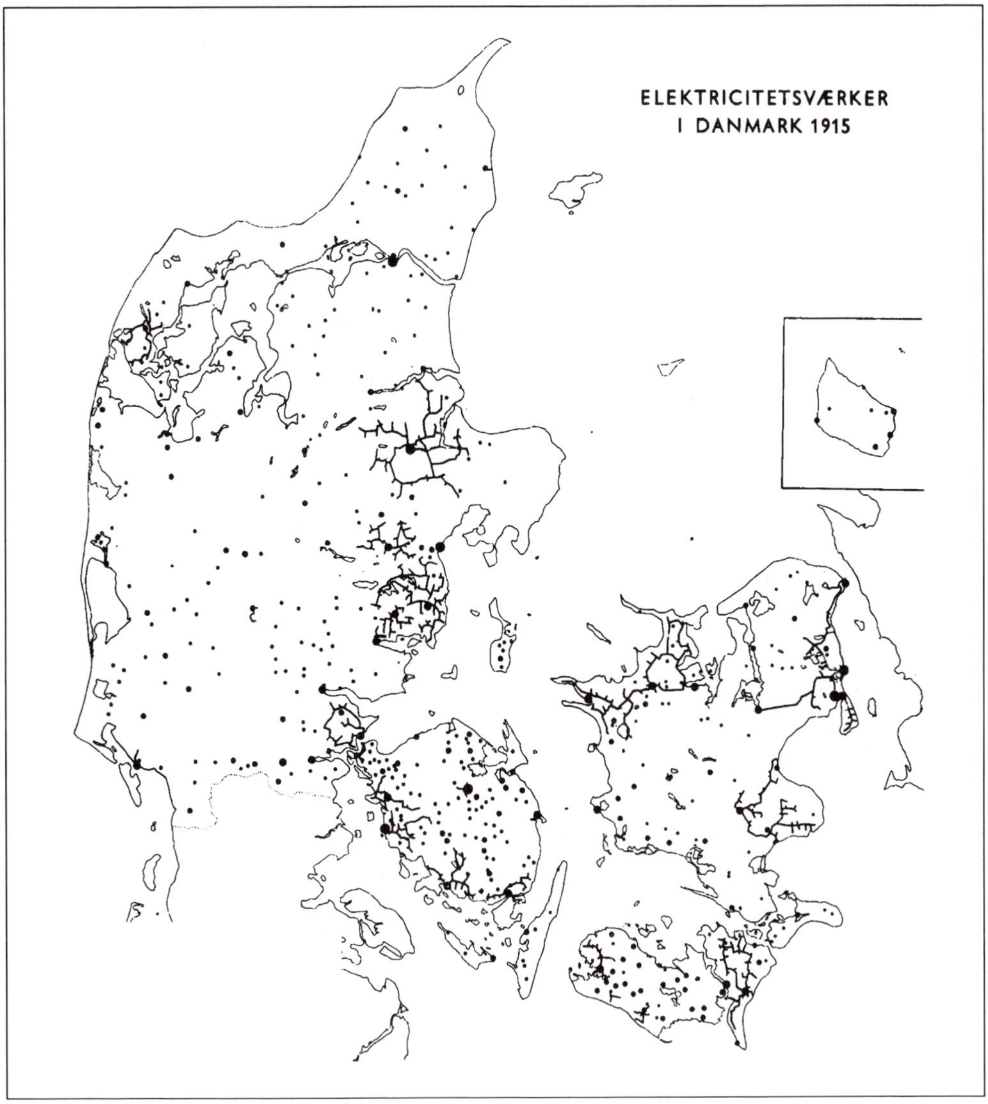

The location of all Danish electricity-generating stations in 1915, ranging in size from below 50 kW (small black dots) to more than 6000 kW (black squares). The drive for centralization is already visible, but the most striking feature is the existence of many small, local stations of less than 50 kW, particularly on Funen and on Jutland where the cooperative movement had its strongholds.

TABLE 3.
Power machines and other machines in Danish agriculture, 1923 (Source: *Det Danske Landbrugs Historie*, 1988).

Power machines	Number	Percentage of farms
Electric motors	39,600	20.4
Kerosene engines	14,700	7.6
Wind mills	16,600	8.6
Stationary steam engines	670	0.3
Traction engines	7,400	3.8
Tractors	2,000	1.0
Other machines		
Threshing machines	111,600	57.6
Grinding machines	45,700	23.6
Milking machines	730	0.4

Soon thereafter the power stations in central Copenhagen changed to AC. In 1915 a seacable combining the networks of NESA and the southern Swedish networks was established, thus allowing the import to Denmark of cheap water-power-generated electricity from Sweden.

Centralization did not proceed as fast as one might have expected, partly because of the great success of the cooperative movement in Denmark. The cooperative movement had put down deep roots in the country after the establishment of hundreds of cooperative stores, dairies, and slaughterhouses during the 1880s and the 1890s. The cooperative movement had inculcated a sense of self-reliance among Danish farmers—and an aversion to large, anonymous companies, led by persons from higher urban society. In the small towns, especially on Funen and Jutland, people preferred small and local power stations, financed and run according to well-known cooperative principles. And an interesting technological development turned out to be in accordance with this way of thinking. The development of the diesel engine at B&W from 1903 to 1906 had resulted in motors with low capital costs, excellent fuel economy, and small maintenance costs. These were suitable as prime movers in local DC generating stations. Small communities, i.e., the many market towns growing up along the ever-denser railway net, got the opportunity to become electrified and still maintain their full independence. Many small towns made use of this opportunity.

1920–1972: A Modern Industrial Society Emerges

On January 1, 1920, the eight-hour working day was introduced by law. In the early spring of the same year, southern Jutland returned to Denmark after a referendum. The country thereby reached its present size, 44,000 km^2. The size of the population has increased from 2.4 million in 1920 to 5.3 million in 1990 and is now almost stationary. Economically the period can be characterized as one of recession through most of the 1920s and 1930s, war economy during the German occupation 1940–1945 and the years immediately after, slow growth in the 1950s, exceptional high growth during the 1960s and early 1970s. Concerning the relative importance of the main occupations, there has been a constant changeover from farming as the most important trade towards industry. The balance point came in 1960 when the total export value of Danish farming products for the first time was less than the export value of all industrial products

Following a referendum in 1972 Denmark joined the European Common Market (EEC).

Agriculture

When World War I broke out in August 1914, the harvest had already started. July had been the hottest anybody could recall, with 23 days of tropical temperatures. At once 20,000 young men were called to arms to secure the Danish borders. Most of the men were recruited from the countryside, and the sudden loss of manpower jeopardized the harvest. Thanks to the widespread mechanization on the farms and an

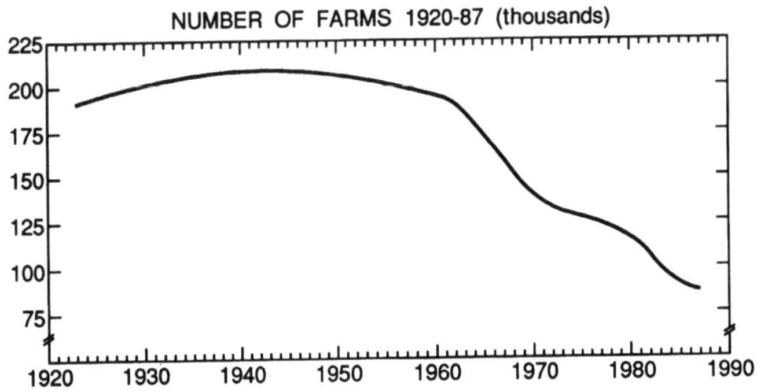

The number of Danish farms, 1920–1987.

enormous work effort from the men who remained at home, the harvest was saved. In a critical situation like this, the farmers were able to manage the work by themselves.

The general picture of Danish agriculture in the 20th century is one of massive mechanization and motorization. In 1919, 733 tractors were imported—600 were the new Fordson. Today the tractor is considered a natural part of any Danish farm. But the picture is less unambiguous than it may look. During World War II, under the German occupation, the number of machines using horses peaked. After 1945 the tractor began to gain momentum on a massive scale. The Marshall Plan in 1948 paved the way for the tiny but very reliable Massey Ferguson, which replaced the horse during the 1950s and 1960s. By 1964 statistically there was a tractor on every farm. In 1985 every farm had more than two tractors at its disposal, but the number of farms had decreased.

Today the typical Danish farm is a highly rationalized industrial plant that can raise several hundred pigs or cows at a time. Agriculture has developed into a coherent technological system, mastered by the farmer with a little help from his wife and kids. During the 1970s, the first personal computers were introduced to manage the economy, fields, and stables. Today the computer is an integral part of modern agriculture. Biological engineering is implemented in the cultivation of species, seeds, and the development of totally new crops.

The number of farms has been drastically reduced since 1960. The farmer has become a technician with a long theoretical education in an agricultural school behind him, before he is allowed to own a farm. This strategy of massive implementation of new technology on the farms has been very successful in raising productivity and total farm output. However, it has also created serious ecological problems with pesticides and fertilizers as the main suspects. Also serious charges from consumer organizations have been levelled against a postulated decline in the quality of many farm products. The Danish parliament has reacted with sporadic attempts to introduce stricter agricultural legislation, so far without too much success.

While the ecological deterioration has been accelerating during the 1980s, many small farms have switched to ecological cultivation of the land. The latest development has shown that bigger farms too are beginning to switch to ecological production on a hi-tech level, which hopefully will enable the farmers to deliver ecological products to the consumers at prices that can compete with those of conventional farm products. In this process the cooperative peasant movement has been a very conservative force. The farmers powerful organizations have bitterly fought every law that has been enacted so far to improve the environment, if the law threatens the income of the farmers.

A more progressive policy seems to be underway among younger farmers today. During the last decade, hundreds of farmers have

Two different high-quality products from the same company. The company, Nielsen & Fisker Ltd., produced Nilfisk vacuum cleaners (from 1926) as well as Nimbus motor cycles (from 1921). In addition to a modern design, both of these products were characterized by long durability. Nilfisk vacuum cleaners are still produced, although the design has changed. 'The bumblebee' (Model 1934) is not produced any more, but it can still be met on Danish highways in the summer!

started to produce electricity from windmills. Also many experiments with biogas systems have taken place on Danish farms using residual dung and mixing it with other forms of waste to produce energy. Today Danish agriculture is at a technological turning point. The farmers who have survived the restructuring of the farms since 1960 today manage a highly efficient technological system, with an enormous output compared to the traditional farm. In return modern farming creates ecological problems and new technological challenges. Will the future farmer be able to combine his Danish farming tradition with new and high environmental standards and thereby give room for a dynamic development in the agricultural sector? Or will the Danish farmer continue in the same conservative direction as his leaders have up till now, with impoverishment and a predictable collapse as the result? This is the technological challenge to the Danish farmer on the threshold of the 21st century.

Crafts and Industry

At the beginning of the period, in 1920, Copenhagen's status as the industrial center of Denmark could not be debated. However, a number of provincial towns, completely dominated by one industry, existed: the sugar beet factories on Lolland-Falster, the cement factories in Aalborg, the train factory in Randers and the paper mills in Silkeborg, Dalum, and Næstved being obvious examples. Similarly, in Helsingør, Svendborg, Nakskov, and Frederikshavn, the local shipyards towered above all other industries in their respective towns, with respect to employment as well as prestige.

In Copenhagen the iron and metal industry together with the textile industry dominated the city, and gave rise to a sex-divided labor market. After the turn of the century, the women's share of total employment had increased dramatically, but in Copenhagen in particular, certain industries became almost completely dominated by female workers. The iron and metal industries, the newspaper and printing occupations were reserved almost exclusively for men. In the breweries and in the shoe factories, women constituted approximately half the labor force, and they outnumbered men in the most labor-intensive trades. In the tobacco and textile industries women constituted about 70 percent of the workers, and in the knitwear factories almost no male workers were employed at all. Mechanization, standardization, and mass production forced their way into these trades after 1920, thus giving rise to unemployment among women, but at the same time, creating new women's jobs as secretaries and shop assistants.

Burmeister and Wain (B&W) was already 'Americanized' during the 1890s, and stood out as a symbol of the new and important role of engineers in modern industry. But it was only after 1920 that civil engineers appeared in Danish industry in great numbers, starting to build up modern production systems after American models. In the decade 1920–1930, one-third of the graduates from the Polyteknisk were employed in private industry. Furthermore, polytechnicians were increasingly being employed as business managers, thus becoming responsible for unifying firm's economical and technical considerations. Attempts to introduce time and motion studies took place, and assembly lines were introduced in some factories, i.e., Ford Motor Company, which in 1919 established an assembly factory in Copenhagen.

The 1920s saw an important international breakthrough for Danish scientific research and for some industries based on scientific research. In medicine, August Krogh was awarded the Nobel Prize in 1920, and together with H.C. Hagedorn, he developed the principles behind the industrial production of insulin, which for a long time, gave a Danish firm (NOVO) a monopoly on the world market. The physicist Niels Bohr was awarded the Nobel Prize in 1922. At the Carlsberg laboratory, an epoch-making analysis of enzymes was carried through. Industrial research laboratories were established in connection with a few companies in the chemical industry, the iron and metal industry, and the food and stimulant industry. This development was supported by the state, for example, by the foundation in 1937 of the Akademiet for de Tekniske Videnskaber ('the Academy of the Technical Sciences') to ensure the industrial utilization of pioneering scientific research.

The 1920s also saw a breakthrough in Danish industrial design. This breakthrough has generally been associated with a handful of famous architects and artists, but it should be emphasized that a necessary prerequisite for the industrial production of the new products was many small industries with a long tradition of excellent craft work. The 'Danish-designed' industrial product was the PH-lamp, a new functionalistic light source developed by the young architect and multi-artist, Poul Henningsen. This epoch-making piece of work was quickly followed by electronic household articles, radios, grammophones, and motorcycles, each product developed by one or more industrial designers trying to unite form and function in a product of high aesthetic value. Production was still almost exclusively directed towards the home market, possibilities for export being only slowly discovered during the 1930s.

The house-building industry was strongly influenced by the functionalistic wave of the 1920s and 1930s, and this holds for design as well as for choice of materials. The Danish cement industry (F.L. Schmidt), one of the few Danish industries able to obtain its raw products from domestic sources, already in the 1930s created a number of standardized building components. It was only after the Second World War, however, that F.L. Schmidt's industrial production of prefabricated components for turnkey element buildings was developed into a proper building industry.

The Second World War turned industrial conditions upside down. Free connections with the world market suddenly disappeared and with them the possibilities of obtaining traditional raw materials for production. Much of Danish industry had to use alternative, domestic materials—often resulting in a 'substitute-product' of poor quality—or they had to use unreliable German raw materials. The substitutes never became popular, and after the war they disappeared with the dismantling of import prohibitions and money restrictions. On the other hand, a few large entrepreneurs ran successful, although unpopular, businesses with the German occupation forces as they became involved in building roads, airfields, and an impressive row of fortresses along the west coast of Jutland. Taken as a whole, there is no doubt, however, that the Danish production apparatus suffered heavily from sabotage during the occupation, and from being without access to repairs and technological developments overseas.

Already in 1947, Danish industry had reached the prewar level of production, with 75 percent of production being sold on the home market. The spectrum of trades was rather broad, and production techniques were mainly craftlike or semiindustrial. Most factories employed fewer than 20 people, only 25 percent of the labor force being employed in factories with more than 100 employees.

The 1950s were a decade of rationalization within Danish industry. A number of steps to improve productivity, often modelled after Taylor's scientific management system, were carried through. 'Continuous flow processes' were introduced in many large factories, as were extensive plans for dividing work processes and for minimizing 'waste time' through time studies and productivity-increasing wage systems. The former craft-dominated production was thoroughly mechanized. From an investment point of view, it was not so much a development towards larger production as towards a more efficient utilization of the production apparatus, the technical and planning content in production being greatly increased. Associated with this development, the number of white-collar workers also increased dramatically.

In addition many companies started emphasizing niche-production of highly specialized industrial products, thereby making them much more competitive on a still more specialized world market. Niche-production fitted into a growing international specialization, a prerequisite being highly educated technicians in sufficient quantities. These were delivered by the Polyteknisk Læreanstalt, which changed its name to the Technical University of Denmark (DTH), and by the many 'teknika' and technical schools all over the country. Danish industry has never been competitive with respect to standardized, mass-produced consumer goods; instead the strategy emphasized quality products, designed for the fulfillment of individual needs. Important industrial enterprises can be found within the iron and metal industry, the electronics industry, the chemical industry, the

Wind electricity-generating station at Vindeby on the Baltic. Still more wind mills are placed in shallow waters where the wind conditions are optimal, the energy production from water-based wind mills being typically 50 percent higher than four similar land-based positions. The Vindeby station consists of 11 450 kW windmills, the distance between each pair being approximately 300 meters. The total production per year is expected to be near 12 million kWh.

furniture industry, and the shipyards, but also within the textile industry, which is mainly located in the middle of Jutland. The textile industry was nearly wiped out by imported, cheap, mass-produced textile products from the Far East in the beginning of the 1970s, but a changeover from labor-intensive to technology-intensive production and a goal-oriented development of high quality, individually designed products brought the crisis to an end around 1980. This constitutes just one example to illustrate how niche-production has been the most dynamical growth area since 1960.

The oil crisis of 1973 was a moment of truth for Danish industry as well as for the rest of Danish society, since the dependence on cheap oil from the Middle East was almost total. Market conditions worsened abruptly, thus stimulating a search for energy-saving techniques and increased productivity within the industry. The crisis favored small and medium-sized companies, reducing 94 private companies with more than 500 employees in 1973 to only 83 in 1981. But the crisis has also accelerated the ongoing movement of industrial companies from east to west Denmark, and from the big cities to the provincial towns. Thus the production of oil from the North Sea has been accompanied by a substantial build-up of a Danish off-shore technology on the west coast of Jutland, making the off-shore industry a new and innovative growth area in Danish industry over the past 20 years. The large-scale development work associated with the increased utilization of renewable energy sources, the wind in particular, should also be mentioned. Today Danish wind-power technology leads the world market, and Danish companies have a yearly export in the order of one billion Dkr. In 1990 more than 2,000 electricity-generating windmills had been erected in Denmark, their total production being approximately 2.5 PJ, which corresponds to 2 percent of Danish electricity consumption.

The shipyard industry also ran into great difficulties after the 1973 oil crisis. The previous lucrative production of supertankers was suddenly brought to a halt, with a massive increase in unemployment as a result. Since 1973 this very market-sensitive trade has gone through a tough transition process, including substantial restructuring, automatization, and product development. Today Denmark has one of the world's largest shipyard industries, announcing full order books for many years to come. Instead of building oil tankers and bulk carriers, the larger shipyards have changed their production, e.g., to advanced container ships designed according to the wishes of the individual customer. The smaller shipyards have experienced a rapid sale in advanced fishing trawlers, which have gradually developed into floating factories, extremely effective in tracing and fishing large shoals of fish.

TABLE 4.
The development of the population, number of cars and railway net, 1920–1980 (Source: *Danmarks Historie*, Vol 9, 1990).

Year	1920	1930	1940	1950	1960	1970	1977
Population (thousands)	3,200	3,551	3,844	4,281	4,585	4,938	5,100
Private cars (thousands)	14	79	117	118	408	1,077	1,375
Trucks etc. (thousands)	4	32	45	61	170	252	269
Km railways	4,328	5,294	4,915	4,815	4,301	2,890	2,498

This development has brought an end to itself, however, due to overfishing and the resultant introduction of fishing quotas in the Atlantic Ocean and the Baltic Sea. Consequently, a number of shipyards are now trying to survive by making another desperate changeover, e.g., to scrapping old trawlers or to producing offshore equipment, windmills, etc.

Infrastructure

Railways, Roads, and Bridges. The Danish railway net continued to grow until 1930 when 5,300 kilometers of railway had been constructed. A considerable part was local railways, not always established for sound economic reasons but rather to please specific local interests. The railway expansion stopped when private cars began to appear in great numbers; railway lines were then closed down almost as fast as they had been built a few decades earlier. As a substitute, the road net was expanded at a feverish pace to keep pace with the exploding car traffic.

The Danish bridge-building era was ushered in when the Social Democratic government during the crisis years, the 1930s, tried to fight the high unemployment rate by implementing various job-creation programs, for example, by building bridges between the different parts of the country. The Limfjord Bridge, the Little Belt Bridge, and the Storstroem Bridge were constructed under the supervision of Danish consulting engineering companies, such as Kampsax, Christiani & Nielsen, Monberg & Thorsen, and Hoejgaard & Schultz. Thus these Danish companies gained considerable expertise in building large bridges, knowledge that was utilized in the following decades when they succeeded in winning bridge-building contracts around the world.

Shipping. Denmark's large foreign trade stimulated the growth of a number of shipping companies with great interests in the shipyard industry. Companies such as DFDS, the East Asiatic Company, and Maersk have found important niches in the world transport market, and they have played a significant role in Denmark, politically as well as economically, throughout this period.

Air Traffic. Although regular Danish air traffic dates back to the twenties, it was only after World War II that it started to play an important role. Scandinavian Airlines System (SAS) was founded in 1950 by fusing the three national Scandinavian air companies into one major company. SAS main airport is situated in Kastrup near Copenhagen, but 11 provincial airports also exist in Denmark. However, many of these will undoubtedly face serious economic problems in a few years when the Big Belt Bridge opens for regular traffic. By then the travelling time between the center of Copenhagen and many towns in Jutland will be almost the same whether by car, train, or airplane.

Telephony. As in many other countries, telephone traffic has grown tremendously since the first Danish telephone company was established in 1879. The number of customers was approximately 100,000 in 1917; in 1990 it is close to 2.5 million. In collaboration with the state-owned Post & Telegraph Company (P&T), the telephone companies are busy modernizing the Danish telesystem, e.g., by exchanging the old coaxial cables for new fiber optic cables.

Radio and TV. Denmark's Radio (DR) started its regular services in 1925, and from 1951 to 1968, black and white television was broadcast as well. Only one TV channel existed, but it tried hard to cover everything from family entertainment to political reports. In the 1980s the media explosion reached Denmark via the many TV satellites, quickly destroying the old structure of the electronic media. A competing TV channel (TV2), partly financed by commercials, is now fully established. In many ways it imitates DR-TV: it delivers almost the same product, but uses far fewer resources. TV2 has been able to take advantage of being a newcomer; it is run by a small

Danish energy consumption, 1950–1988, and how it is satisfied by the four main sources, oil, coal, natural gas, and renewable energy. It should be observed that energy consumption has scarcely risen since 1970. The natural gas comes exclusively from Danish sources in the North Sea. From 1991 Denmark has also been self-sufficient in oil.

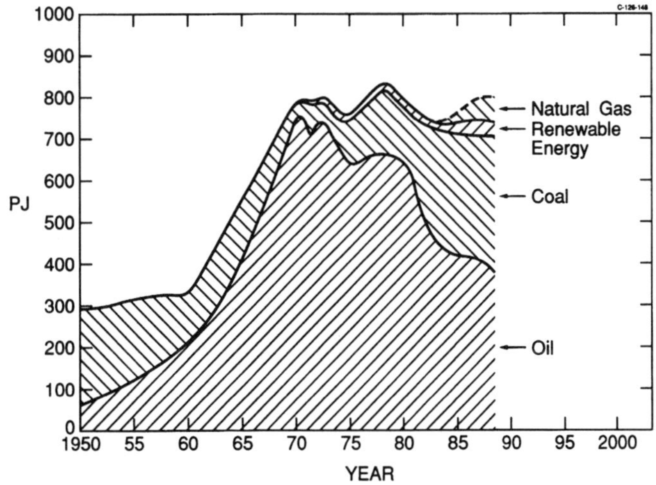

administration, and its managers are allowed to use the newest technology much more efficiently than their colleagues at DR-TV.

Electrification. The centralizing tendencies already visible within the electricity supply system before the First World War, gathered momentum after 1920 with increased collaboration between neighboring central stations. The final goal was the establishment of a standardized electricity system, based on mutually connected coal- or oil-burning central stations, which would generate high-voltage alternating current with the most up-to-date dynamos or steam turbines. Thus the electricity supply could be secured even if a breakdown occurred somewhere in the system, and furthermore, the overall load factor could be improved, which again would mean an improved economy. A network covering the whole country would have the further advantage that customers throughout Denmark could use the same lamps, motors, electric household articles, etc.

More and more DC-stations were closed down during the 1930s and 1940s, and in 1954 all the AC-networks in east Denmark were fused into one large network, Kraftimport, which for economic reasons staked much on the import of Swedish water-power-generated electricity. A corresponding network, ELSAM, in 1957 materialized in west Denmark. It was connected with the large Norwegian network. In the beginning of the 1960s, local enclaves with small DC-stations were still in existence, but their days were numbered.

Just after the 1973 energy crisis, the Danish government proposed to reduce Denmark's almost total dependence on imported oil by substituting for some of the oil-burning power stations nuclear power stations. A strong public reaction against this plan and an overwhelming public support for an alternative plan to reduce energy consumption and to increase the use of renewable energy forced the government to give up its attempts to introduce nuclear power to Denmark.

Although the different parties in Folketinget, the Danish parliament, disagree on many details, the great majority have agreed on an energy program that in only 17 years has made Denmark almost self-sufficient in energy. The means to accomplish this have been: (1) Incentives to stimulate the insulation of all houses to obtain extensive savings in the energy consumed for heating. (2) Reduction of the electricity demand through the development of new and energy-saving technology, for example, in refrigerators, freezers, electric ovens, etc. (3) Enhanced search for oil and natural gas in the Danish sector of the North Sea, and the construction of production facilities in places with sufficient reserves to allow for economically feasible production. (4) The establishment of an extensive distribution network for natural gas (in 1973 natural gas was not used anywhere in Denmark), and the establishment of oil transmission lines from the oil fields in the North Sea to the oil refineries in Denmark. (5) A considerable research and development effort in renewable energy—wind, sun, and biomass in particular—so that these sources in the future will contribute a much larger fraction of Denmark's total energy supply.

Outlook

Just like the rest of the industrialized world, Denmark today faces serious problems that are intimately connected with modern technological developments. How to solve these problems will no doubt be one of the biggest societal issues to be debated in Denmark over the coming years. Shall we take it upon ourselves to create those technical solutions that are in accordance with our traditions and fulfill our needs? Or shall we chance it? Accept that we cannot do anything ourselves against ongoing developments, and consequently let solutions come to us from abroad? This essential discussion of the proper direction of future technological development may soon be intense in the field of energy consumption, CO_2-emissions, etc. Shall we accept the high-tech solution: a swift changeover to nuclear power on a large scale? Or shall we try to follow the route suggested by some of the 'grassroot movements': a determined reduction of our country's energy consumption and giving top priority to research and development in renewable energy sources? We think our country has developed the societal and technological potential to be able to realize each of these alternatives over the next 50 years. The choice is essentially a political question, and the final choice will no doubt have far-reaching consequence for the development of Danish society in the 21st century. . . .

The material in this chapter has to a large extent been extracted from a number of voluminous research based works surveying the history of Denmark from a political, economical, social or cultural point of view. At present research works in Danish (contextual) history of technology are virtually non-existent, but several research projects are now underway.

Further Reading

Danmarks Historie ("The History of Denmark").

Det Danske Landbrugs Historie ("The History of Danish Agriculture").

Dansk Mejeribrug ("The Danish Dairy").

Industriens Historie ("The History of Industry").

Dansk Industri efter 1870 ("The Danish Industry after 1870").

Håndværkets Kulturhistorie ("The Cultural History of the Handicrafts").

Københavns Industrialisering ("The Industrialization of Copenhagen").

H.P. Clausen, S. Ellehøj, and S. Mørch (eds.). *Danmarks Historie*, Vol 1–9. København, 1991.

Vol 1–4. Claus Bjørn (ed.). *Det Danske Landbrugs Historie*. København, 1988.

Claud Bjørn (ed.). *Dansk Mejeribrug*. Odense, 1982.

Axel Nielsen (ed.). *Industriens Historie*, Vol 1–3. København, 1944.

Hans Chr. Johansen (ed.). *Dansk Industri efter 1870*, Vol 1–3. Odense, 1988.

Vagn Dybdahl m.fl. (ed.). *Håndværkets Kulturhistorie*, Vol 1–4. København, 1984.

Ole Hyldtoft. *Københavns industrialisering*. Herning, 1984.

Technological Change in Finland

Timo Myllyntaus

The Axe People and Swidden Cultivation

The Finns are a small people who have for centuries lived at the crossroads where various larger nations meet and have thus been influenced by different cultures. Traditionally, Finland has been seen as a halfway house between East and West.

While moving with their ethnic relatives, other Baltic-Finnic peoples, from Siberia westward to the Baltic Sea, the Finns adopted a great many tools and skills from the Indo-Europeans, not least from the Balts who were their southern neighbors. From the ancestors of the Latvians and Lithuanians, the Finns borrowed, for example, the axe, *kirves* (*kirvis* in Lithuanian), which became their principal tool for over two millennia. The Finns continued their journey northeast to the coasts of the present Gulf of Finland and Gulf of Bothnia. Using the axe and other borrowed technology, they settled down there and pushed northwards the former inhabitants of the region, the nomadic Lapps with their reindeer.

In Medieval Europe, the Roman Catholic and Greek Orthodox churches were religious and political rivals. It was the Finns' misfortune to live between these two blocs, which sought to expanded their influence northward. From the mid-11th century, the Christian Russian rulers sent their troops from Novgorod to subjugate and baptize the heathen Finns. A century later, the Swedes became increasingly interested in overseas areas to the east. By organizing crusades, the first of which took place in 1155, the Kingdom of Sweden succeeded in seizing a foothold in the southwestern part of the Finns' territory. Sweden later extended its conquests and managed to keep Finland proper under its power for over 650 years.

The country where the Finns settled was not an ideal habitat. It had a cold climate and numerous fens, which gave the name, Finland, to the whole country. There were few natural resources; the most valuable proved to be abundant forests, passable soil for agriculture, and a labyrinth of lake and river systems. Since the Stone Age, the Finns' destiny has been bound with the forest. The forest has often given them the means for living, provided indispensable energy sources, and sheltered them against enemies and even against the oppression of their own authorities. Hence, the Finns did not only live in the forests, but they also earned their living from them.

In the preindustrial period, nearly all buildings, bridges, fences, ships, carts, furniture, and most tools were made of wood and primarily fashioned with axes. The Finns also used the axe to cut clearings in the forests. Following their hunting routes, Finnish settlers gradually advanced northward into the backwoods practicing swidden cultivation or slashing and burning, which was the oldest type of agriculture in Finland, dating back to the Stone Age. Originally, that archaic method of grain-growing required few other tools: primarily hoes, sickles, wooden handspikes, and fire.

Until the end of the Middle Ages, most Finns lived in the coastal zone, and only in the 16th century was central Finland permanently settled. Swidden cultivation and population growth basically supported each other. That type of grain-growing normally supported a family year round, but it was also labor intensive. A peasant practicing swidden cultivation needed a workforce to fell timber, burn clearings, sow, reap crops, etc. He tended to have work associates—kinsmen or strangers. The peasant's junior partners were called axemen, *kirvesmiehet,* who often lived in his household, whether they were related to him or not.

Women also cleared woodland as shown here at Eno parish near Joensuu. The fire had to be constantly attended, and the heaps of burning wood had to be rolled across the clearing. The ashes fell on the ground and fertilized it.

Source: I. K. Inha, *La Finlande pittoresque,* Helsingfors 1896.

Swidden cultivation was practiced in a rotational manner; a peasant family used to return to their earlier plot after 15 to 40 years. The Finns developed different techniques of slashing and burn-clearing. Two of the most popular methods are worth mentioning. *Lehtokaski,* the oldest type, meant slashing a 30- to 40-year-old broadleaf forest, burning it the following year, and then sowing barley or rye in the fertile ash. After that, the peasants grew oats for one or two years in the burn-beaten area until they abandoned it to be forested naturally by broadleaf trees, such as birch and alder.

The Savo tribe in eastern Finland specialized in slashing and burning. For it, swiddening was not only an agricultural method but also a way of life. In the late Middle Ages, the tribe adopted a new technique from the east, probably from the Russians, and developed new variations of it. In the basic form of this *huuhtakaski* method, peasants made a clearing in the coniferous forest by barking standing larger trees and slashing smaller trees and bushes early in the spring. Then in summer, they burned the slashed trees, but left any standing timber to dry for a year. After that they felled and burned the larger trunks, too. The peasants of Savo sowed a special sort of grain, wildwood rye, in the autumn and reaped the harvest the following fall, i. e., in the third year after starting cultivation. When it was successful, *huuhtakaski* yielded an enormous harvest compared to an ordinary crop from field cultivation, but the same plot could only be used two or three times before it had to be left to fallow for several decades. This method allowed the Finns to profitably utilize even rather poor soils for grain-growing from one to four times a century.

The kings of Sweden were well aware of the efficiency with which the Finns used swidden cultivation for settling uninhabited woodlands. Because Swedish peasants were reluctant to inhabit the forested province of Värmland along the southern border between Sweden and Norway, the Swedish government transferred people from eastern Finland to settle that area. Finnish swiddening peasants were also well represented among the people that Sweden sent to Delaware, its 17th-century American colony. The American cultural geographers Terry Jordan and Matti Kaups claim that the Finns of Delaware created the agrotechnical basis for the conquest of the American West. Compared to other European immigrants, the Finns had superior experience in and peerless techniques for clearing uninhabited woodlands. They were able to make long hunting trips in deep forests and could support themselves in the wilderness in other ways, too. In addition, the Finnish immigrants had an extraordinary ability to communicate with the Indians and learn from the Indians' experience. They were among the first to adopt from the Indians the use of corn, turkeys, maple syrup, and medical herbs, for example.

Elias Brenner chose a man hewing balks with an axe to symbolize the trades of 18th century Finland when he drew his famous copper engraving with the country's coat of arms. The engraving was published in the book *Suecia antiqua et hodierna* in 1716.

Source: Erik Dahlberg, *Suecia antiqua et hodierna*, Stockholm 1716.

Swiddening depletes great amounts of forest. When it was extended to large areas, it threatened other uses of timber. It conflicted with the interests of ironworks, sawmills, shipyards, etc. Furthermore, due to the rapid population growth in Finland in the 18th century, swidden cultivation exhausted itself. When the same plot was slashed and burned too frequently, overcultivation prevented the soil from regaining its fertility. Harvests diminished, and consequently in the 19th century, famines became more severe. There were economic and political pressures to give up swiddening in favor of field cultivation and fertilizing the soil with manure. Swidden cultivation continued in Finland until the late 19th century when Finnish agriculture began to be commercialized and to change to dairy production.

The Proto-industrial Era, 1500–1840

Until the early 20th century, Finland was an overwhelmingly agricultural country. Both rural and urban dwellers made most of the goods they needed by themselves or with the help of their servants. Country villages had a barter economy, and their inhabitants purchased few things from towns: primarily salt, iron, and handicrafts, such as tools, kitchen utensils, textiles, and later some beverages and tobacco. They also needed money to pay taxes. Because the use of money was predominantly limited to the transactions of the bourgeoisie and the government, there was not much purchasing power for industrial products in the country.

The early stirrings of industrialization involved manufacturing for export. In the beginning, these activities cannot be called export industries. The term *proto-industrial production*, meaning a preliminary stage of manufacturing, is more appropriate. Proto-industry was generally a seasonal activity; peasants and agricultural workers used the quiet seasons of farming to weave, sew, hew logs, make woodwork, and forge metal for the merchants in the towns. Working usually at home or in the nearby forests, they used their own tools. The products were intended for sale in markets outside the producers' own region; they were often transported to Stockholm by merchants' barks or by peasant-sailors' boats.

A traditional proto-industrial product was tar. Tar-burning was an old industry in Prussia even in the Middle Ages. It spread around the Baltic Sea, flourished in the forested regions, and died out slowly. Tar exports from Finland began in the late 16th century. The demand for Finnish tar and its distillate, pitch, began to increase markedly in the early 17th century. The big seafaring countries needed tar to coat their vessels and protect ropes against rot. It was also used to protect wooden buildings, as a lubricant, and for many other purposes. The great coniferous forests had mostly disappeared from the British Isles and Central Europe, which raised the commercial value of Finnish sawn timber and other wood-based

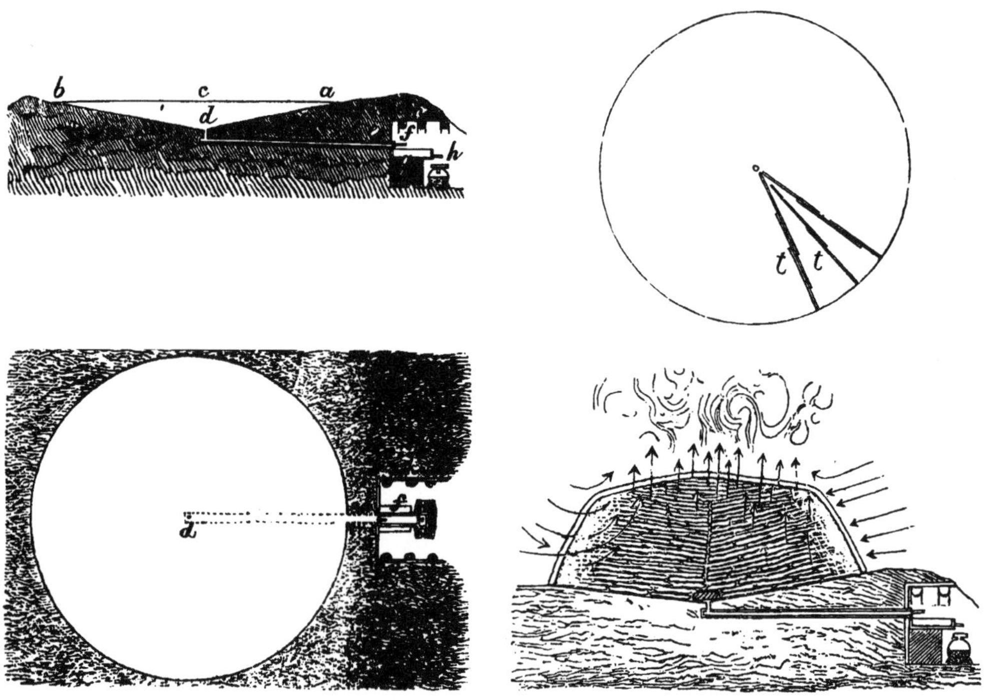

The Finnish tar-pit of the 19th century was a coniform hollow with tightly trodden clay walls. Resinous billets were piled radially in the pit. The heap was covered with moss and soil. During the burning, a pipe led tar from the very bottom of the pit to a nearby trough and then to a barrel. *Source:* Erik Daglberg, *Suecia antiqua et hadierna*, Stockholm 1716.

products. From western Finland, great quantities of tar and pitch were shipped to Stockholm and then reexported to Britain, Holland, and France. Eastern Finland tended to export its tar and pitch directly to the Western markets.

Customers considered the quality of Finnish brands, known as "Stockholm tar" and "Viborg" or "Crown tar," higher than that of their American and Russian competitors. To produce tar was a long process. It began when peasants barked the standing pines with axes to make them dry out. Some years later, they felled trunks and chopped logs before smouldering them in tar-burning pits. Peasants transported tar barrels by boat through lakes and rivers from inland forests to the seaports and sold them to merchants. Eastern, central, and northwestern Finland were the most significant regions of tar production in the Swedish-Finnish kingdom supplying most of its total exports. Among the 17th-century exports from the kingdom, tar ranked third after iron and copper. In the early 19th century, Finland exported about four times more tar than Sweden. Consequently, tar and pitch were prominent Finnish export items from the 17th to the mid-19th centuries. During that period, Finland was the world's greatest tar producer.

As a result of the 1808–1809 war between Sweden and Russia, Finland was ceded to the Russian Empire as an autonomous grand duchy, a political status it had never had under Swedish rule. For the Finnish economy, the separation from Sweden caused no abrupt change. In many respects, free trade continued between them for several years, and Swedish bank notes were lawful money along with Russian roubles until the Finnish monetary reform of 1840. The change in Finland's political status involved an economic tradeoff; it opened promising new economic perspectives in the east, while gradually hampering trading opportunities with Sweden.

Most Finnish proto-industrial production was based on the use of wood as a raw material and source of energy. A complementary energy source was waterpower. The watermill was invented in the Middle East about 100 B.C. It took twelve centuries before this prime mover was introduced into Finland. The first document concerning a watermill dates from 1352 when it was mentioned that the archbishop of Turku owned a water-powered flour mill at the rapids of Halistenkoski on the bank of the river Aura. The new technology spread slowly in the inland provinces. However, from the 16th century on, the diffusion of both horizontal and vertical watermills substan-

Oulu at the mouth of the river Oulujoki was an important export harbor for Finnish tar. After tar-boats had managed to escape from the hard embrace of the Merikoski rapids, they glided into smooth water at Toppila Sound, where the tar-depot was situated. With ropes and cranes, the boats' crew lifted heavy tar barrels onto the quay.

Source: I. K. Inha, *La Finlande pittoresque*, Helsingfors 1896.

The picture on the left describes the mechanism of a traditional frame-saw with several blades. With thc driving bar and the crankshaft, the saw was directly connected to the axle of the waterwheel. Gears became common in Finnish sawmills in the early 19th century. The picture on the right presents the typical construction of a small sawmill in the late 19th century. Although many vital parts of the sawing machinery were made of iron, the legs of the sawframe were still built of wood.

Source: Artturi Käpy, *Sahateknologia I, Sahakoneet,* Helsinki 1927, p. 4.

tially speeded up. Manors, peasants, and merchants set up flour mills all over the country.

Watermills replaced manual work and the muscle power of horses in grinding grain. The Finns gradually learned to apply watermills to other purposes as well. The first water-powered sawmills were set up in Finland in the mid-16th century. Technology for them came from the West, primarily from Sweden. These plants were modest wooden buildings with a wooden vertical waterwheel directly coupled to a wooden sawframe. Almost the only iron part was a coarse, thick blade that sawed trunks with slow, vertical reciprocating movements. Sawing a log into planks with that kind of apparatus took hours. Such equipment was not a serious competitor to the axe-hewing and hand-sawing that continued in Finland until the end of the 19th century.

Although saw-milling was a small industry, the authorities attempted to restrict its growth. They considered it a threat to forest resources that they wanted to save for more valuable purposes, such as the production of iron and shipbuilding. The wood famine of Central Europe and the local problems of Swedish ironworks were reflected in the forest policy of the kingdom. In the 18th century, the government's restrictive measures dampened the growth of saw-milling in Swedish-ruled Finland. In the treaties of 1721 and 1743, Sweden had ceded some of southeastern Finland to Russia, whose tsars as a rule pursued more liberal forest policies than did the Swedish kings. In "Old Finland," meaning the Finnish territories under Russian rule, saw-milling therefore developed rapidly in the 18th century. Russian entrepreneurs, merchants, and noblemen participated in the sawmill business in Finland. In relative terms, their activities climaxed in the 1830s and 1840s. In eastern Finland in 1835, Russians owned 16 sawmills that employed about 44 percent of the sawmill workers in that region. At the time, Russians also influenced the technology and organization of Finnish saw-milling. For example, the Finnish term for a forest working site, *savotta*, is derived from the Russian word *zavod,* meaning hauling.

After the waterwheel and one-blade frame, the next major innovation in proto-industrial saw-milling was the so-called Dutch fine-blade saw. This saw worked more efficiently and with less energy. It also reduced the waste of raw material. It was faster than a saw with one coarse blade because several thin blades could be fixed in one frame, and one waterwheel could sometimes drive two parallel frames. According to contemporary estimates, the older sawmills with a coarse blade could saw from five to seven trunks a day, whereas a fine-blade saw with one frame could handle up to 20 trunks daily.

Dutch-made fine blades were first introduced in sawmills near Riga and Narva in the mid-17th century. From there the new technology spread via Ingria and the Karelian Isthmus along the northern coast of the Gulf of Finland westward in the early 18th century. Eventually the innovation reached as far as Ostrobothnia. From Finland it passed over the sea to Norrland in Sweden proper. The first fine-blade sawmill in Norrland was erected at Kramfors in 1744 by Christopher Kramm, a sawmill builder from Finland, who had fled the attacking Russian troops during the war of 1742–1743. In southwestern Sweden and Norway, this innovation came into general use almost a century later than on the Baltic coasts. In the Nordic countries, the fine-blade sawframe is a rare example of an innovation spreading from the east to the west. Most innovations have moved in the opposite direction.

A more ordinary example of the transfer of technology is shipbuilding, which the Finns learned by working in Swedish shipyards in the early 17th century. Some shipyards were later established in Finland, too; two Swedish entrepreneurs set up an important dockyard at Pietarsaari in 1666, and the Swedish Crown founded another near Kokkola in 1675. The craft of building seafaring ships was diffused among the peasants of Ostrobothnia who were willing to hew and carve pine logs during the quiet seasons of agriculture. Because the Finnish workforce and the local pines were cheap, ships were easy to sell in Stockholm or abroad for a profit. From the late 17th to the end of the 18th centuries, shipbuilding was probably the most important industry in Finland. The country exported about 2,500 vessels between 1727 and 1807. Shipbuilding was concentrated in Ostrobothnia, which earned much more export income from ships than from tar.

The iron industry was very important in making Sweden a European Great Power in the 17th century. Due to worries about the supply of charcoal, the Swedish government was willing to use the larger forest resources of Finland for iron production. It set up the first ironworks in Finland at Mustio in 1616. Later both private entrepreneurs and the government established more ironworks, but no promising iron deposits were found in the country. The ironworks on the west coast had to import most of their rocky iron ore from Sweden proper. Ironworks in east Finland used indigenous lake or bog ore that was hoisted by traditional peasant methods. The Finns did not borrow only the term *rautaruukki*, ironworks, from Sweden *(järnbruk),* but they also adopted architectural styles and imported equipment for their works from across the Gulf of Bothnia. Over the centuries, peculiar industrial milieus developed around these ironworks.

After the treaty of 1721 following the Great Northern War, a glorious period began for the iron industry of the Swedish-Finnish kingdom. It was, first of all, caused by soaring demand for iron in Britain. In 1740, it has been estimated that Sweden-Finland accounted for 40 percent of the world's iron production. However, at best, Finland's contribution to the kingdom's total output was just three percent. By the end of the Swedish period, eight blast furnaces and 24 bar iron hammers were working in the country.

Except for tar-burning, saw-milling, ship-

From the 17th century, Finnish ironworks used water-powered hammers to forge iron. The forge of the Strömfors ironworks was still working at the beginning of the 20th century.

Source: Suomen paeri- ja puutavaralehti 29 (1937), p. 339.

building, and the ironworks, other industries were present in Finland only on a very small scale. It was only in the 1640s that the first notable glass factory was founded in the Swedish-Finnish kingdom. The founder of this factory in Stockholm, Melchior Jung, had a son who became the trailblazer of the Finnish glass industry. In his factory, set up at Uusikaupunki in 1681, Jung Junior began to produce glass for windows. By the end of the 17th century, glass windows had become general in that town and at the provincial center, Turku, and they slowly spread to the inland areas. The number of glass factories increased, especially in the 1790s. Besides window glass, they also began to manufacture other items under the supervision of foreign, mainly German, masters.

Still in the 1830s, most dwellings in central and eastern Finland were smoke-cottages—meaning log huts without chimneys or glass windows. Nevertheless, a new architecture and a new living style were spreading from the southwestern parts of the country to those regions from the 18th century. In the countryside, the houses of the gentry and priests were first supplied with bright glass windows and tiled stoves, originally a Dutch innovation adopted into Finland from Sweden proper. In the peasant houses, glass windows were smaller, and stoves were built without Dutch tiles. Nevertheless, more important than the decorated tiles was that peasants, too, adopted stoves with billet-saving flues and chimneys.

Stoves with fairly efficient flues were developed in Sweden in the 18th century. This work carried out by Count Carl Johan Cronstedt and others, was motivated by the desire to rationalize space-heating, reduce the effort of handling billets, and preserve precious forest resources. From the 17th century, the Swedish-Finnish elite became increasingly interested in Central-European economic doctrines. Among other things, it also adopted a new view about the use of natural resources; this way of thinking later became known as environmental consciousness. To avoid the overuse and exhaustion of timber resources, the government intervened with various measures on swidden cultivation, tar-burning, saw-milling, the iron industry, and even space-heating.

The Academy of Turku, the cradle of Finnish university education, was founded in 1640. At the same time, the Lutheran Church was putting more effort into educating the populace and teaching it to read. Books were needed to raise educational standards, and printing books was impossible without paper. Primarily to provide the preconditions for basic education, Bishop Johannes Gezelius, Senior founded a papermill in Pohja parish in 1667 to manufacture printing paper from linen and woolen rags. During the Swedish period, the Finnish paper industry remained very modest and worked only for the domestic market. One reason for its small output was that the level of education remained low despite ambitious intentions. Although an estimated 50 to 75 percent of the Finns could read in 1800, less than four percent could write. However, those who were unable to read, probably did not miss much; hardly any printed secular writings in Finnish were available—just a few religious texts, such as psalmbooks and catechisms.

Gründerzeit—the Dawn of Modern Industrialization, 1840–1890

No epochal event happened in Finnish society around 1840. Nevertheless, various moves toward modernization took place. The Finnish government, *Senate*, as it was called, formulated its economic policy in a new way. It decided to loosen its ties with Sweden and strengthen trade relations with Russia, while also seeking to increase the economic independence of the grand duchy. Consequently, the Senate made marked changes in its monetary, financial, and foreign trade policies. In 1842 it also set up a new central civil service department, *Manufaktuurijohtokunta,* or the Board of Manufacturing, to promote industrialization. These changes hardly received quick responses from business life; the upturn in the economy can mainly be attributed to other factors: the opening of new trade relations, the setting up of new kinds of factories, and the introduction of up-to-date machinery. Examples of modern, mechanized factories were cotton mills, engineering workshops, and a papermill with a British paper-making machine. Not only were new machines introduced in the 1840s, so were unprecedented prime movers, namely, the reciprocating steam engine and a prototype of the hydroturbine.

In 1828, the Scottish machine builder James Finlayson founded the first cotton mill in the grand duchy with Russian credit. The mill started its development towards becoming a large factory after its ownership was transferred to the Baltic German businessmen Carl Samuel Nottbeck and Georg Rauch. By the 1850s, the Finlayson cotton mill had become the largest cotton factory in the Nordic countries due to its voluminous exports to Russia. Finlayson's specialties included an iron waterwheel with a diameter of 27 feet and a capacity of 50 horsepower. This wheel was built by Fiskars, the first Finnish engineering workshop, in the late 1830s. Contemporaries regarded it as one of the largest waterwheels in Europe. In the late 19th century, the mill was expanded, and several new buildings were erected on its premises. In Scandinavia, there was nothing comparable to the mill's Plevna, which was a huge factory hall supplied with the most modern machinery.

Between 1840 and 1890, half-a-dozen other cotton mills were set up in the grand duchy. At the end of that period, the four largest cotton mills employed together about 4,000 workers. The small number and large size of the factories distinguished the Finnish cotton industry from its Scandinavian counterparts.

Prime movers were important articles for Fiskars, whose engineering workshop was established in 1837. Besides iron waterwheels, it manufactured steam engines for ships and factories under the management of English and Swedish technicians. In Finland, steam power was first introduced in lake transport. In 1833, a steam paddler, the *Ilmarinen,* owned by two Karelian sawmill owners, was launched; this tugboat was powered with a Russian-made steam engine of 34 hp. The earliest stationary steam engine, which was also foreign made, was put into operation in the Havi candle factory near Viipuri in 1840. Fiskars manufactured its first stationary steam engine for the Littoinen cloth factory in 1844. Its steam engines soon became famous in neighboring countries. Between 1838 and 1853, the workshop exported half of its steam engines to the Baltic provinces of Russia.

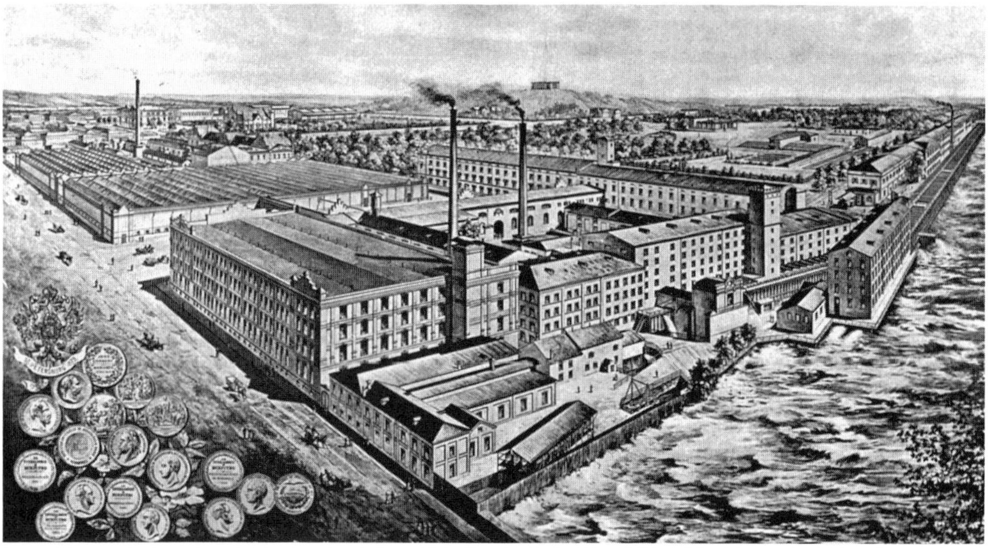

An aerial view of the Finlayson cotton mill drawn at the turn of the century, before aerial photography was available. The larger of two one-story buildings with a 'saw-blade roof' is Plevna. That weaving hall had a capacity of 28.500 cubic meters and was 139 meters long and 39 meters wide. The smaller building on the left was also a weaving hall.

Source: Finlands handel, sjöfart och industri, Tammerfors, Ed. Julius Hirn, Helsingfors 1908, p. 34.

Reciprocating steam engines never gained the upper hand in Finnish industry; in many sectors they served as a reserve capacity in case of drought. The hegemony of the waterwheel was crushed by a different prime mover, the hydroturbine. This water-power motor was invented by the Frenchmen Benoit Fourneyron in 1827. His invention did not become generally known in Finland until press reports years later. Starting in 1847, Fiskars produced more simple water-powered prime movers, so-called Scottish reaction wheels. Fiskars' technicians learned the technology to produce them from Sweden. It was only in 1854 that the first proper hydroturbines were put into operation in Finland. The engineering workshop of Finlayson then manufactured two Fourneyron turbines for the factory's own use. Nevertheless, the diffusion of the hydroturbine was fairly rapid; by 1875 it had become the most important prime mover in Finnish manufacturing measured by its proportion of the total capacity of industrial motive power. In terms of generated energy, the hydroturbine preserved its leading position up to the mid-1960s when it was replaced by the steam turbine. Hence, contrary to the teaching of the standard textbooks, in Finland the Industrial Revolution was set up and primarily carried through by water power—not by steam power.

A peculiarity of the period 1840–1890 was that a surprisingly large number of the prime movers and working machines of industry were made by domestic workshops. Foreign machines were imported, but many were soon imitated by Finnish manufacturers. Such a versatility of engineering products was a conspicuous achievement for an agrarian country that was just beginning its industrialization.

The mechanical pulping machine is an example of a successful adoption of a foreign innovation. In the early 1840s, the German weaver Friedrich Keller invented a machine that mechanically pulped wooden blocks into the raw material of paper. In Finland, the first experimental pulp mill was put into operation in 1860. However, large-scale production started with a German Keller-Voelter pulping machine and a Finnish Jonval hydroturbine at Tampere in 1866. Due to the high quality of ground woodpulp, the founder of the mill, Fredrik Idestam, was in 1867 awarded a bronze medal in the Paris World Exhibition. The success encouraged Idestam to build a larger mill at Nokia in the late 1860s. The factory needed efficient equipment for pulping blocks of spruce. Therefore, he designed a new type of wood-grinding machine with Herman Kauffmann, an engineer of German descent who had moved from Gothenburg to Tampere. By 1875, 12 mechanical pulp mills had been founded in the grand duchy, and many of them were supplied with so-called Kauffmann-Tampella grinding machines.

Mechanical wood-based pulp could not replace all rug pulp in papermaking. The next step was the production of chemical pulp from wooden chips that started on a commercial scale in Finland in the mid-1880s. The introduction of three key innovations, the papermaking machine, mechanical woodpulp, and chemical woodpulp, revolutionized the whole paper industry. Formerly, Finnish papermills were hampered by a chronic shortage of rags and the low productivity of handicraft methods. Three brand new industries—mechanical pulping, sulphite, and sulphate pulping—were set up to utilize up-to-date technological innovations, and the paper industry developed from a shabby, small-scale manufacturing to the most modern and fastest growing industrial sector in the grand duchy. A lack of raw materials no longer hindered the development of paper production, which found enormous markets in the Russian Empire. The forest had again proved to be the basis of the Finns' welfare.

The mercantilist economic policy of the old Swedish type was continued in Finland up to the end of the Crimean War (1853–1856), when the reforming tsar, Alexander II, decided to liberalize economic life in the grand duchy. One of the most decisive changes took place in saw-milling; the restrictive regulations and tight forest policy were abolished, and free enterprise was allowed. In 1857 the Senate decided to allow the use of steam-driven sawmills, and in 1861, it rejected all restrictions concerning annual quotas for sawn timber. In this respect, Finland followed the policy of its neighbors with a lag of one or two decades. The first steam-driven sawmill was established in Estonia in 1838, and some years later in the regions of St. Petersburg and Riga, and at Tunadal in Sweden in 1849. The earliest three steam-driven sawmills were commissioned in Finland in 1860.

Owing to extraordinary export demand, Finnish saw-milling experienced a hectic boom in the early 1870s; in seven years, the output grew by 290 percent. Due to large investments in new production capacity, steam-powered mills surpassed watermills in terms of output in 1874. The boom attracted Norwegian entrepreneurs and workers to Fin-

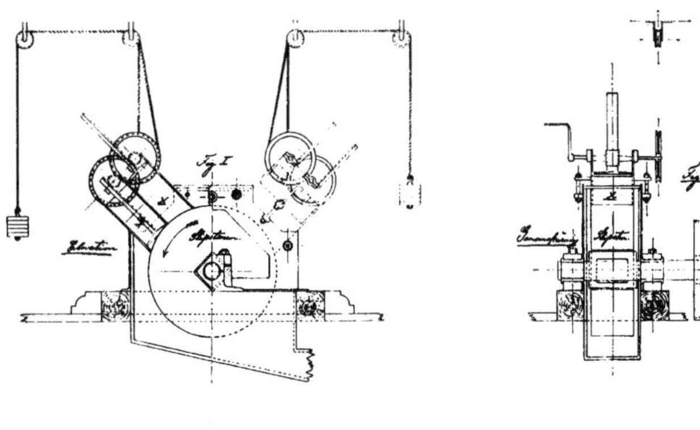

In 1869 Fredrik Idestam and Hermann Kauffmann patented their grinding mill with a horizontal axle. The machine is an early example of successful Finnish innovation and indigenous machine design.

Source: Lars G. von Bonsdorff, *Nokia aktiebolag 1865–1965*, Helsingfors 1965, p. 81.

land. Their participation in the timber business helped modernize timber floating and the Finnish sawmill industry.

In the 19th century, saw-milling was Finland's largest and most prominent industry. Sawn timber was the major staple product exported to the Western markets; the significance of other Western-oriented export products, such as tar and later butter, was much smaller. An advantage of saw-milling was that export returns were not limited to a small part of the population. They were distributed fairly broadly, both regionally and socially; the peasants as sellers of timber received their share. Lumberjacks, timber floaters, sawmill workers, and stevedores got their shares as wages, and entrepreneurs gained their profits. Thus, the export income increased domestic demand and promoted the expansion of the monetary economy. Saw-milling was undoubtedly a key in promoting the growth and modernization of the agrarian Finnish economy.

In addition to saw-milling, there were three other large industries in the grand duchy: textiles, iron, and paper. Their modernization and growth were greatly based on the demand of the Russian market. These industries benefited because industrialization started more slowly in the Russian Empire than in Finland. Between 1840 and 1890, the grand duchy was a rare phenomenon; for a backward agrarian area on the northern periphery of Europe, it had a surprisingly versatile industrial output and some fairly large factories, although the domestic demand for manufactured goods remained modest. The pivotal role of exports was a peculiarity in the first phase of the grand duchy's industrialization. As a result, foreign trade accounted for a strikingly large part of its gross national product.

The Gründerzeit was not only characterized by nascent modern industries but also by various investments in the infrastructure. One achievement of the early 19th-century Senate's mercantilist policies was the construction of the Saimaa Canal connecting the country's largest lake with the Gulf of Finland. The canal was built in 1845–1856, and it later became an important outlet for eastern Finnish products to the markets of St. Petersburg and Western Europe. The new border of 1944 cut traffic on the canal, but as a result of a Soviet-Finnish agreement, the Saimaa Canal was rebuilt in 1962–1968. Thereafter, it has managed to compete with highways and railroads in the transport of bulky goods.

The liberal Senate, which came to power after the Crimean War, preferred railroads to canals, and during its reign, "iron horses" carried the first passengers from Helsinki to Hämeenlinna in 1862. The opening of the Helsinki–St. Petersburg railroad in 1870 was

The Saimaa Canal of 59 kilometers connects Lake Saimaa to the Gulf of Finland. It was a valuable achievement of the mercantilist policy of the Senate. Although the canal was narrow, it was an important means of communication for transporting timber and other goods by steamer from Savo and north Karelia to the Baltic Sea. In many places near Lauritsala, the canal was cut through granite. *Source:* I. K. Inha, *La Finlande pittoresque*, Helsingfors 1896.

economically more important event for the 1.75 million Finns because it provided relatively fast and cheap transportation for exports to the empire's capital. By that date, St. Petersburg with its 670,000 inhabitants had become a European metropolis and the fifth largest city on the Continent.

Industrial Breakthrough amid Turbulence, 1890–1945

Despite advances in many sectors, the general industrial base of Finland at the end of the Gründerzeit was weak. In 1890, only 52,000 persons, or 6.7 percent of the economically active population, were working in industry and handicrafts. The following 55 years would not seem to have been favorable for industrialization. First, the period was politically turbulent. In the 1890s, the Russian authorities began to russify Finland; opponents of this policy were arrested and sent to Siberia. Censorship was intensified. The oppression was relaxed from 1905 to 1908, but then it worsened again. When World War I broke out in July 1914, the Russian government declared a state of war and restricted civil rights in the grand duchy. Finland's declaration of political independence (1917) was followed by civil war in the spring of 1918. During the whole interwar period of the 1920s and 1930s, internal political tension continued between the right and the left, and the Soviet Union and Finland were not able to normalize their mutual relations. The Winter War of 1939–1940 was a disaster for Finland, but it united the divided nation. The Continuation War of 1941–1944 was another hard trial for the Finns. The truce of Moscow in 1944 did not end the fighting because the Finnish army had to drive the Germans from Lapland. As a result of these three wars (1939–1945), Finland suffered 90,000 casualties, lost Petsamo and a part of Karelia—together about 11 percent of its territory—to the Soviet Union, and Lapland was badly damaged.

Second, economic conditions were not very promising. The 1890s started with harvest failure and a depression. An even longer recession opened with the next decade. World War I interrupted Finland's foreign trade with its major western partners, Britain and Germany. The civil war weakened the opportunities for a postwar revival. Due to the turmoil in Russia, Finland's foreign trade with its eastern neighbor ceased after the October Revolution of 1917. The loss of a prominent trading partner was a hard blow that the compensating markets in the West could not fully offset. In the 1930s, soon after the country had overcome the lean years of the Great Depression, World War II brought even more severe hardships.

Third, in the 1890s, it became evident that poorly industrialized Finland could not follow the most advanced countries in technological developments. Many of its factories then proved to be too old fashioned. For example, small, traditional Finnish ironworks could not compete with the large modern plants of Ukraine and southern Russia. Consequently, instead of exporting iron to Russia, Finland gradually began to import it from the empire. The products of the Finnish engineering works were also gradually replaced by Western manufactured goods both in the Russian and home markets. In addition, the textile industry was compelled to change from an exporting sector to a home-market sector. After World War I, Finnish exports continued to lose their versatility, and the monocultural trend was intensified in Finnish manufacturing.

It is a great paradox that despite all the unfavorable and even depressing features of the period 1890–1945, industrial development was nonetheless extraordinarily rapid. During that distressed era, an industrial breakthrough actually took place in Finland. By 1939 the country had managed to narrow the gap with the leading industrialized economies of the West. After the civil war, Finland quickly recovered from its wartime setbacks, and its industrial output accelerated before the slump of 1929–1933. During the 1930s, the growth rate of its industrial production was the fourth swiftest in Europe.

Again timber was a cornerstone of the Finns' success despite all adversities. The wood-processing industries kept developing, and thanks to their products, Finland's terms of trade remained favorable. It was fortunate that wood-processed products proved to be good export items; demand for them in the world market was rising, while oversupply did not become a prevailing phenomenon.

Up to the outbreak of the slump of the 1930s, saw-milling remained the most sizeable industry in the country. By 1913 Finland had become the second largest exporter of coniferous wood, logs, and sawn timber in the world. For several years after 1924, it was the leading exporter of sawn timber. The country did not, however, attain that place due to its technological superiority. Its success was based on huge forest resources and the ability to produce and export tremendous amounts of timber during a period of widespread instability and unpredictable fluctuations in markets.

Finland also had other thriving mechanical wood-processing industries. The most successful were those that took advantage of the wide birch forests. Finland's exceptionally good birch groves were a result of the extensive swidden cultivation of the preceding decades. World War I interrupted the expansion of the birch spool industry. In the 1930s, Finland nevertheless ranked first as the producer of wooden spools for textile factories. An estimated 80–85 percent of the spools sold in the world market were Finnish made.

Another prosperous industry in the interwar period was plywood, a board manufactured by gluing thin veneers crosswise together. The Finns learned how to make plywood from the Estonians just before the Great War. In the 1920s, the production and exports of birch plywood grew on average 20 percent annually. Due to the fast expansion of its output, interwar Finland was the world's largest producer of plywood.

Mechanization was an essential trend in saw-milling and in other wood-processing industries from 1890 to 1945. Finnish engineering firms surpassed their Swedish competitors at the turn of the century by launching framesaws with a continuous feeding of logs, an improvement invented by Julius Nielsen, a Norwegian engineer living in Finland. At the same time, the electrification of sawmills accelerated, and due to electric power transmission, the layout of the mills could be rationalized, and new kinds of sawing, hoisting, and transporting machines were introduced. During the interwar period, the artificial drying of sawn timber became more general, and attempts were made to improve the quality of products.

These novelties caused no epochal change in saw-milling. Despite various investments, productivity in the manufacturing of standard sawn timber did not rise substantially. To maintain their viability, Finnish sawmills paid more attention to utilizing their wastes and developing remunerative by-products. With special saws and low-cost female labor, some of the wood waste was converted into export articles, such as broomsticks, boxstaves, and plaster laths. Formerly, wood waste had been burned or dumped into the sea or lakes. By introducing steam engines and especially turbogenerators, part of the waste could now be turned into useful energy. From the turn of the century, sawmills expanded their electricity generation, selling part of it to neighboring factories or local utilities, which distributed electric power to households and other consumers.

The most profitable way for sawmills to utilize their wood waste was by producing chemical pulp. The Swedish engineer Alvar Münzing invented a method of producing dark, firm sulphate pulp from coniferous chips. His method was excellent for making outstanding, heavy-duty wrapping paper from sawmill waste. The problem was that Münzing carefully hid his chemical secrets. Through industrial espionage, the Finnish Walkiakoski paper mill managed to obtain the hidden know-how by persuading a Swedish engineer to sell process specifications in 1895. Inside Finland, the secrets of the method were diffused relatively quickly, and seven other "kraft-pulp" mills were set up between 1903 and 1920. Unlike groundwood pulp and sulphite pulp production, this new industry was only loosely integrated with the domestic paper industry. This is partly attributed to the fact that sulphate cellulose was not used for papermaking in Finland except in the Walkiakoski paper mill. Nearly the whole output was exported to Western Europe. In contrast, sulphate pulp mills became closely linked with the sawmill industry. In fact, the sulphate pulp industry developed into a satellite industry for sawmills, which found pulp production a good way of profiting from their refuse.

The Finnish paper mills preferred to use sulphite pulp because their main products were newsprint and other bleached papers. The machinery in sulphite pulp mills developed in several major phases. The central device was a pulp cooker, the average size of which grew by a factor of ten between 1885 and 1920. This growth took place owing to the change from simple Swedish bowl cookers to larger, more sophisticated German cooker models.

In mechanical pulping, Finnish mills aimed at making ever more pulp per machine and per kilowatt hour. The factories worked in two or three shifts, and mill owners kept on investing in more efficient grinding machines. For example, from the 1890s, they started to replace their Finnish-made machinery with new American warm-grinding machines directly coupled with rapidly revolving Francis hydroturbines that gave a twofold capacity compared to their predecessors. The new American machines provided the inspiration for improving the older machines, and consequently, technological change in the whole industry accelerated markedly.

Groundwood pulp mills did not only improve their quantitative production facilities, but also raised their value-added factor by refining part of their pulp into cardboard. Furthermore, many of these mills became vertically integrated with domestic paper factories. Similarly, sulphite pulp mills were established mainly to supply a semi-manufactured material for paper factories in the country. As a result, all three types of production units—mechanical grinding, sulphite pulping, and paper-making—worked as closely integrated parts of fairly large corporations.

Improvements in the production of softwood pulp had great effects in paper-making. Due to the introduction of sulphite pulp, rag pulp could be replaced entirely. While there were enormous amounts of coniferous timber available for pulping, shortages of raw fiber material no longer restricted the expansion of paper production. As a result, better wood pulp improved the quality of Finnish paper and raised the value-added factor of the country's paper production.

In late 19th-century Finland, two main types of contemporary paper-making machines were in use. With the simple one-cylinder type, the so-called Yankee machine, mills produced cheap wrapping paper for the European markets. In contrast, mills provided with more sophisticated Fourdrinier machines consisting of several cylinders mainly manufactured newsprint and other printing paper sorts for the Russian market. The Finnish paper industry possessed a versatile machinery of a high technical standard on the eve of World War I. Therefore, all its main items managed to compete in the Western markets in the interwar years when there were no more opportunities to export pulp or paper to the east, i.e., to the Soviet Union. Between 1921 and 1937, exports of paper quadrupled, whereas shipments of chemical pulp increased by a factor of ten. Hence, the pulp cookers and chemical pulping methods were the pivotal innovations of the breakthrough period, and chemical pulping led the country's industrial growth.

The success of the wood-processing industries spurred the Finnish economy to export-led growth and helped the less thriving sectors, such as metallurgy and especially engineering, to survive by offering them, too, a chance to modernize. However, nothing could save the old-fashioned Finnish ironworks. The total exports of iron and iron products fell from 22,600 tons in 1897 to an average of

In the interwar period, the axe and the cross-cut saw were still the tools used to fell trees. Lumberjacks worked in frost and deep snow.

Source: Keksintöjen kirja, Puu I, Ed. Martti Levón, Porvoo 1933, p. 97.

4,000 tons between 1906 and 1912. The last charcoal furnace was closed in 1928.

The Finnish metallurgical industry declined for two reasons. First, Finnish ironworks were too small and technologically obsolete. Most of the plants could not adopt the new iron and steel technology that was advancing quickly in the major industrial countries. They generally lacked both the know-how and the capital resources for modernization. Between 1879 and 1913, eight small Siemens-Martin furnaces were set up in Finland. Due to technological problems, however, steel production, grew very slowly. The development only speeded up in the early 1890s. Second, Finland lacked the favorable prerequisites for a modern metallurgical industry, such as good rocky iron ore and coal deposits. It had only a few, poor rocky ore deposits, lake and bog ore, and firewood. This discouraged foreign entrepreneurs and investors from setting up plants and transferring up-to-date technology to the grand duchy.

During World War I, a new turn took place in the Finnish basic metal industry due to the introduction of the electric furnace. Elektrometallurgiska Ab began to produce foundry pig iron at their Vuoksenniska and Nokia melting plants in 1916–1917. The necessary technology was bought from Sweden where the first so-called Elektrometall furnaces had been built in 1910. Because the war interrupted the importation of fossil fuels, the electric furnaces contributed to Finland's self-sufficiency: the industry could use indigenous charcoal as a raw material and hydroelectricity as indispensable energy. In 1937 the Finnish electrometallurgical industry took a big step forward. The Imatra ironworks of Oy Vuoksenniska Ab brought into operation one of the world's most modern electric furnaces. The capacity of the furnace was 100,000 tons of pig iron per annum. Oy Vuoksenniska was for long the only company in the country producing iron from ore; in 1959 the electric furnace was closed, but the factory has continued to operate under the name of the Imatra Steelworks. During the war, metallurgical production expanded, and, for example, electric steel furnaces were commissioned by only three companies by 1945.

On the eve of World War I, a new mine was opened to utilize an exceptionally rich copper deposit at Outokumpu in eastern Finland. In 1925 the government decided to purchase the company's whole capital stock and expand its production considerably. A decade later, Outokumpu Oy set up a modern, fully electrified copper-melting plant. In 1936 it was the third largest copper furnace in the world. Six years later, Outokumpu also built an electrolytic copper mill at Pori to refine its raw copper. By the late 1930s, Outokumpu Oy had become an important exporting company.

During the last decades of the grand duchy, the development of the Finnish engineering industry compensated for the downward trend in metallurgy. At the turn of the century, there were many ambitious engineering projects under work. The engineering companies produced a wide range of electrical equipment and all kinds of prime movers, except steam turbines. Some of them were even able to start manufacturing complete paper-making machines for Finnish mills and submarines for the Russian navy. During the interwar period, the engineering industry managed to enter only a few new fields, such as the manufacturing of elevators, radio receivers, electric cables, and electric bulbs, while it ceased producing some other products, such as paper-making machines. Interwar Finland exported engineering products in only modest amounts, but due to protectionist import tariffs, engineering companies were able to expand their share of the home market. For example, domestic hydroturbines largely replaced imported ones.

Finnish industry and towns electrified relatively rapidly. From the spring of 1882, the Finlayson cotton mill was lit with one of the earliest Edison lighting installations in Europe. The first urban block station was opened in Helsinki in 1884. By 1914 a third of the total motive power installed in industry was electrified, and at least one electricity supply utility was working in 38 towns. In the interwar period, electrification proceeded swiftly. Measured by the degree of electrification of industrial mechanical drive, Finland had become one of the leading countries by 1938. Although rural electrification advanced slowly in the central and northern provinces, by the outbreak of the Winter War of 1939, almost half of the rural households and over nine out of ten urban households were wired for electricity.

Rapid electrification was based on heavy investments in both hydroelectric and thermal turbogenerator plants. In 1921 Parliament, against the will of the government, decided to build a state-owned large hydroelectric power plant at the Imatra rapids. The plant was put in

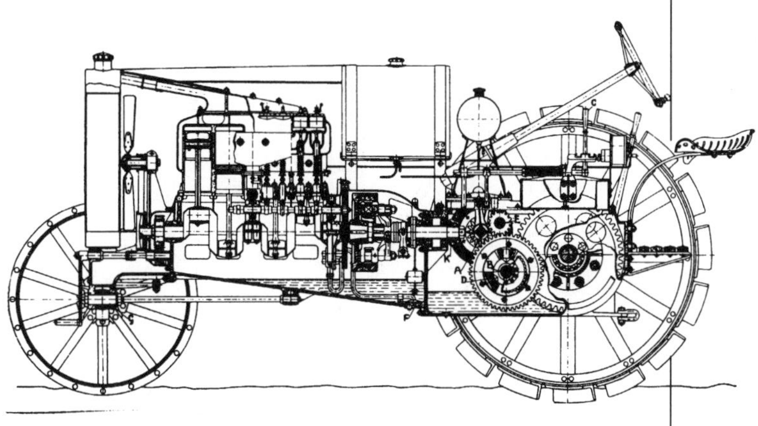

During World War I, the engineering works Turun Rautateollisuus ja Vaunutehdas Oy decided to start making tractors. Designing a tractor of 30 hp took three years, and within that time, a Finnish engineer made a study tour of the USA. In 1919, the first Kullervo tractor was completed. In six years, 300 tractors of this type were manufactured.

Source: Keksintöjen kirja, Maatalous ja ravintoaineet, Porvoo 1939.

operation in the late 1920s, and it succeeded in delivering electric power for the industrial and population centers of southern Finland. The generation capacity of the plant increased from 56 MW to 125 MW between 1929 and 1937.

A sign of Finland's late industrialization is that the most important railroad-building boom took place between 1883 and 1904 when the length of the railroad network grew by about 100 kilometers per year. The following noteworthy upswing was realized between 1925 and 1929. Then the average annual increase was 140 kilometers. Between 1890 and 1939, the length of railroad lines tripled from 1,910 km to 6,120 km. Most Finnish railborne transport has always been under the government's control; on the eve of the Winter War, the State Railroads owned 95 percent of the lines. Due to the cessions of territory to the Soviet Union, a quarter of the network was lost in the early 1940s. In the postwar period, the operated railroad lines have increased from 4,460 km to 5,870 km despite the tremendous expansion of highways and the rapid growth in the number of automobiles.

As during the Great War, the imports of solid fuels substantially decreased, and the imports of liquid fuels nearly stopped, during World War II. Germany supplied Finland with indispensable solid fossil fuels, but it could deliver little kerosene and gasoline. Some solid fossil fuels could be replaced with indigenous fuels, primarily with billets, but supplying fuel for automobiles was a major problem in wartime Finland. The whole shortage could not be filled with puusprii, "woodspirit," alcohol distilled from the waste broth of cellulose. Therefore, wood-gas generators were installed many civilian automobiles, buses, and trucks, while gasoline and naphtha were reserved for military vehicles. Thanks to indigenous energy resources, wood and waterpower, Finland avoided the most severe repercussions of the wartime energy crisis.

Especially in the Winter War, Finland lacked up-to-date military arms, especially antitank weapons. In this bad situation, the Finns innovated a simple grenade using mostly indigenous, wood-based components. They poured tar, "woodspirit," and a few drops of kerosene and gasoline into an empty vodka bottle. In battle, soldiers smashed this "Molotov's cocktail" against the ventilation

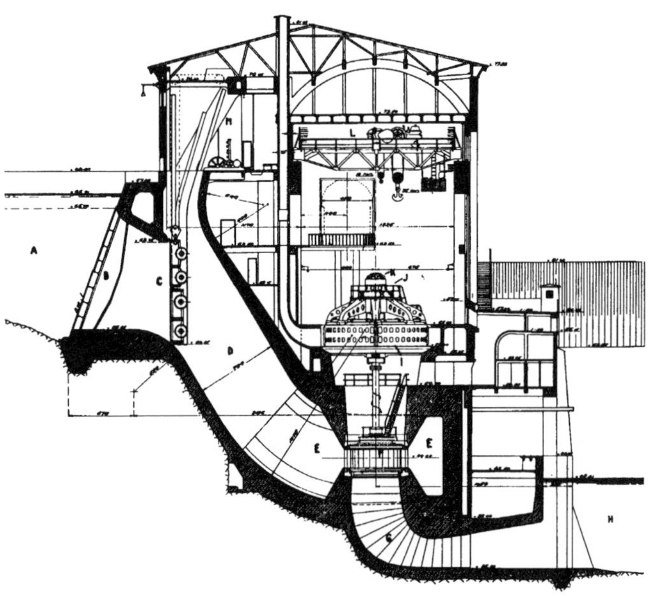

The Imatra hydroelectric power plant was the basis for the Finnish national grid. This state-owned plant was commissioned in 1929; its prime movers were four Francis hydroturbines. The generation capacity of the plant grew from 56 MW to 125 MW between 1929 and 1937.

Source: Iso tietosanakirja, Helsinki 1930–1939.

The sawing hall at the Warkaus sawmill of Ahlström Oy was well lighted by electric bulbs and fluorescent lamps in the late 1950s. Automatization had decreased the workforce since the late 19th century.

Source: Ahlström Oy, Warkaus.

grating of an enemy tank, which generally set the engine on fire and sometimes caused the tank's fuel system to explode.

Economic Growth and a Technological Leap, 1945–1992

Hard times did not end when fighting ceased in spring 1945. The Finns had to face difficult tasks: the settlement of 480,000 evacuated Karelians, the reconstruction of ruined Lapland, the transformation of the production structure for peacetime purposes, the revival of the economy despite currency shortages, and the payment of war reparations to the Soviet Union. The Allies forced Finland to pay the same amount of war indemnities as Hungary and Romania, countries with at least twice Finland's population. Not only the quantity of war reparations, which annually took between one and six percent of Finland's GNP during the years 1944–1952, but their quality caused problems. The Soviet Union demanded that only one-third of the indemnities be wood-processed goods. The Finns would have preferred to use more of their traditional export items because the production of standard sawn timber, pulp, and paper did not cause technological difficulties for them.

The Finns were astonished that almost two-thirds of the reparation shipments had to be in the form of engineering products, such as ships, locomotives, and prime movers. Furthermore, Finland had to deliver sophisticated machines to the Soviet Union, some of which had never before been produced in Finland. The Finnish government subsidized heavy investments in both state-owned and private engineering firms, which attempted to acquire the necessary technological know-how by nearly any method. For example, foreign machines were copied without license or permission.

In the early years of the Cold War, the U.S. authorities were suspicious of Finland. The

Truman government limited the exports of machinery to Finland because it believed that any technology might be reshipped to the Soviet Union. Finland, therefore, had special difficulties in importing up-to-date equipment for reconstruction and new investments even if it had had the currency to buy it. Although its stocks of currency remained modest, the country refused to receive Marshall aid due to its possible political repercussions.

For some years no cars, for example, were imported. There was a danger that the country would be left technologically behind the rest of Europe. The whole economy—not only its foreign trade—was tightly regulated. In the protected environment and with government subsidies, the engineering industry, however, got a special boost. Many new products began to be manufactured under foreign licenses or on the basis of domestic research and development. Since the late 1950s when the government began to liberalize foreign trade, a surprising number of nascent industries or modernized branches of old industries proved to be competitive, first in the Scandinavian market and then in other Western markets. For example, the cable industry grew rapidly and was export-oriented. Kone Oy, which headed the Finnish elevator industry, gradually developed into a multinational company. In the postwar years, four companies began to produce paper-making machines, and after a hard competition, Finland became a leader in this field in the 1970s and 1980s.

After the war reparations were paid, Finnish shipyards continued to supply various vessels to the Soviet Union for over three decades. Icebreakers were one of their specialties; half of the working icebreakers in the world today are Finnish made. Before the end of the bilateral Soviet-Finnish clearing trade in 1990, the most significant delivery of ves-

Finnish paper machine technology has greatly developed within the last 30 years. During that period, the wire width of the largest machines manufactured in Finland has doubled from about 200 inches (5 m) to 400 inches (10 m); design speed has doubled from some 2,600 feet per minute (800 m/min) to 5,200 fpm (1,600 m/min); modern automation has tripled production capacity from some 80,000 to 250,000 tons per year. The Valmet fine-paper machine with a paper width of 370 inches (9.4 m) is among recent exports; it went into operation at the Nordland paper factory in Norway in June 1991.

Source: Valmet Paper Machinery, Jyväskylä.

A Finnish shipyard built two nuclear-powered, shallow-draft polar icebreakers for the Soviet firm Murmansk Shipping Company. These ships, the *Taymyr* and the *Vaygach*, were put into service in 1989–1990. In open water, their speed is 18.5 knots.

Source: Masa-Yards Inc, Helsinki New Shipyard.

sels to the Soviet Union consisted of two nuclear-powered icebreakers with a 50,000-horsepower capacity each. The Soviets supplied nuclear reactors, boilers, steam turbines, and some auxiliary machines. The design of the ships, their framework, and various equipment were Finnish. The total value of Finnish contributions to these icebreakers accounted for FIM 2,000 million. When the trade agreement was signed in 1980, the order was the largest delivery that any single Finnish company had ever received. For the Western market, Finnish shipyards produced large ferries, luxurious cruise liners, and oil rigs until they encountered serious economic setbacks in the late 1980s.

During the postwar reconstruction, the wood-processing industries had to earn hard currency for the country. In this respect, they managed very well, partly because Finland's terms of trade developed so favorably. In the late 1950s, the wood-processing industries started a strong investment program and soon restored their position as the leading manufacturing sector. Internal changes took place within this sector: in the early 1950s, sawn timber was the country's most important export item followed by chemical pulp. By the end of the decade, newsprint, however, had become the major export product. In the 1980s, lwc-paper, or light-coated printing paper for journals, surpassed all other products in the wood-processing sector. The contribution of this sector to total Finnish exports decreased from 85 percent in 1950 to about 40 percent in the late 1970s and remained on that level throughout the 1980s. For several decades, the diversification of the production structure has been a major target of Finnish industrial policy. Nevertheless, the wood-processing industries are still the leading and most influential sector of the economy, which continues to lean on the "green gold," the largest per-capita timber resources in Europe.

Until the mid-1940s, the chemical industry was underdeveloped in Finland. Under the protection of the regulated economy, more efforts were made to promote this sector. In

Modern security systems are examples of Finnish high technology. This Abloy control lock works with a remote infrared key.
Source: Metra Oy, the Joensuu factory.

1948 the state-owned Neste Oy was set up to work in the oil business, and nine years later, it commissioned its first oil refinery. The company received a monopoly on the import of raw oil. Since the 1960s, Neste Oy has expanded its activities in petrochemical production and in the international oil business.

Another state-owned company, Kemira Oy, started to produce fertilizers in the early 1920s, but in the postwar period, it has rapidly grown and entered many other fields. Various other firms produce artificial fibers, industrial chemicals, detergents, paints, plastics, and medicines. For 40 years, the chemical industry has been the most rapidly growing industry in Finland.

After the war, the shortage of electric power was a handicap for the country's economic revival. To compensate for the lost power plants of Karelia and to fill new energy needs, the government launched an extensive plan for building new power plants. The emphasis was in developing the hydropower resources of the northern rivers Oulujoki, Iijoki, and Kemijoki. In the beginning, most equipment was imported, but gradually more Finnish made hydroturbines, generators, and transformers were ordered. By the mid-1960s, the most promising rapids had been put to use in power generation. Subsequently, the rising demand for electricity was met by building conventional thermal power stations. In 1977 and 1981, the two units of the Loviisa nuclear power plant were put in operation. Their nuclear reactors and steam turbines were Soviet made, whereas the rest of the equipment was mostly Finnish or Western. The Swedish company AB ASEA-ATOM dispatched two other units as turn-key deliveries to the Olkiluoto nuclear power plant. These plants were commissioned in 1979 and 1982 respectively. As a result, the total capacity of Finland's four nuclear power plants rose to 2,160 MW. In 1988 nuclear power plants contributed 35 percent of the country's total electricity supply. Only seven countries had a higher proportion of nuclear electricity.

During the first three postwar decades, rural electrification was completed. Today Finland's consumption of electricity is among the highest in the world. The extensive use of electricity has been identified as a basis of Nordic welfare. Nevertheless, even before the oil crisis of 1973–1974, a national goal has been to avoid lavish energy spending. For example, Finland has long been the top country in applying the combined generation of heat and electricity by backpressure steam turbines in thermal power plants. Both industrial plants and district heating systems utilized this technology, whose thermal efficiency could rise to 85 percent. Furthermore after considerable efforts, the insulation and air conditioning of buildings have improved during the recent decades. Triple or quadruple glazing in dwellings and offices has also promoted energy conservation.

Finland only developed an industrialized society in the postwar period. Socially this meant among other things massive movement from the country to the towns. Hardly any European country has experienced such rapid urbanization and such a profound change of occupational structure. The postwar period has also seen a tremendous technological change. Factories have been totally reequipped with up-to-date technology, and manufacturing has become very capital and energy intensive. Households have replaced their wood-burning stoves and outhouses. Most dwellings now have electric, central, or district heating and are equipped with running water and a lavatory. During the 1980s, it was popular to supply dwellings with a fully equipped bathroom and an electrically heated sauna. In many respects, Finland has managed to catch up with the other Nordic countries. The gap in living standards has substantially narrowed between the Finns and the other Scandinavians during the present generation. Within the past 50 years, the country managed to leap from semidevelopment to mature industrialization with high technological standards.

Conclusions

The economic and technological transformation of Finland from a sparsely populated, swiddening area to a welfare state of five million inhabitants has not to any great extent been based on indigenous inventions. The country has bred few famous inventors. In the natural sciences, only one Finn has received a Nobel prize. In 1945 Artturi Ilmari Virtanen got the Nobel prize for chemistry for his contributions to biochemistry and especially for developing a method of preservating silage for cattle. Almost as great an inventor was Eric Tigerstedt (1886–1925), an engineer educated in Germany. He made important contributions to electric acoustics and electronics. For example, he invented sound film in 1915. Patents were granted to more than 400 of his applications. He did not, however, get international recognition for his contributions although his inventions were utilized in the commercial production of electronics.

Finnish inventions have often involved making small improvements to foreign machinery; many of them have been related to redesigning or to adapting foreign technology to local circumstances. Finnish engineers have improved many components of paper-making machines and icebreakers, for instance. Examples of more simple products adapted to the cold climate were the first stainless steel ice skates and the winter tires for cars developed by Nokia Oy in 1936.

A secret of the country's brisk economic development has been the Finns' eagerness to adopt foreign innovations and their ability to be quick learners. The Finns might be considered conservative in many respects, but their attitude to technology has often been progressive. They have been interested in technology that they consider a key prerequisite to a better life. The harsh and capricious climate has probably pressed the Finns to look for security and for reliable means of living. Once someone has proved to them that a new technology is superior to the old one, the Finns have been ready to follow the new and have tended to adopt an innovation within a short period. Thus, although the transfer of a foreign innovation to Finland has sometimes taken a relatively long time, it has generally been diffused quickly around the country.

During the period of industrialization, the transfer of foreign technology to Finland has been massive, and on surprisingly many occasions, it has also been successful. By their efforts and achievements, the Finns have been shown to have a practical and technologically minded character, although they have—at least up to recent times—lacked the precious inclination to make original and epochal inventions and innovations.

Further Reading

Sven-Erik Åström, *From Tar to Timber. Studies in Northeast European Forest Exploita-*

tion and Foreign Trade, 1660–1860. Commentationes Humanarum Litterarum 85 (Helsinki: The Finnish Society of Sciences and Letters, 1988).

Terry G. Jordan and Matti Kaups, *The American Backwoods Frontier. An Ethnic and Ecological Interpretation* (Chicago: Johns Hopkins University Press, 1989).

Timo Myllyntaus, "Hydro- and Thermal Power in Finnish Industry in the Nineteenth and Twentieth Centuries," *Scandinavian Journal of History* 8 (1983), 2, pp. 110–118.

Timo Myllyntaus, *Finnish Industry in Transition, 1885–1920, Responding to Technological Challenges* (Tammisaari: The Museum of Technology, 1989).

Timo Myllyntaus, *Electrifying Finland. The Transfer of a New Technology into a Late Industrialising Economy* (London: Macmillan & ETLA, 1991).

Jussi Raumolin (ed.), "Special Issue on Swidden Cultivation," *Journal of the Finnish Anthropological Society* 12 (1987), 4, pp. 185-265.

Raimo Salokangas, et al. (eds.), *Waterpower in Finland* (Tampere: Finnish Water Power Association, 1969).

Patterns of Technological Development in Norway

Gunnar Nerheim

When looking back on the course of technological development in the Norwegian economy, two generalizations can be made:

- technology was and is strongly attached to the exploitation of natural resources that played a dominant role in the different periods.
- Norwegians have shown great skill in transferring and adapting foreign technologies to the needs and opportunities occurring in the environment of the national and regional economies.

In the following pages, we aim to give examples that can illustrate the process of technological development in Norway from the 16th century until the present. We will take a closer look at several different types of economic activities and the strategies chosen by different participants in a changing national and international environment. In the course of this exercise, we will show that there are, though differences exist, some common traits during those 500 years.

The Exploitation of Abundant Norwegian Natural Resources by the Danish King

From 1536 to 1814, Norway was a province of Denmark, ruled by the Danish king and aristocracy. During those 300 years, the Crown and Danish capital were instrumental in exploiting some of Norway's natural resources. Despite this the Norwegian economy experienced continuous growth, like many European countries in the early modern period. The general increase in European demand for raw materials stimulated the production and export of goods, such as lumber, fish products, and metals.

Export of Norwegian dried cod had been important since the late Middle Ages, and the trade was controlled by the Hanseatic League in Bergen. Cod was caught by local fishermen along the northern coast. Until the 16th century, the east coast of Sweden was the center of the seasonal herring fisheries. In the 17th century, the herring disappeared from the traditional fishing grounds and instead migrated to the western coast of Norway. The herring

Painting by J. Frich, ca. 1840, of Fossum Jernverk. The Fossum ironworks by Skien was established by the Danish king in 1543. It was planned and, during its first decades, operated by Saxon experts.

Source: The Norwegian Museum of Science and Technology

fisheries gave local farmers an opportunity to earn cash income during the off season in agriculture. An increasing number of local merchants entered the fish-export trade, buying catches of cod and herring from local fishermen, processing the fish, and exporting it not only to Germany and the Netherlands, as in the Middle Ages, but also to towns on the Baltic and the Mediterranean.

From the beginning of the 16th century, brisk European demand for lumber led to the rise of the sawmill industry. The water-powered sawmill was introduced, consisting of a crude one-bladed saw. This innovation, being more dependent on suitable streams and the easy availability of logs than on skill and capital, diffused rapidly in the fjords adjacent to the rich timberlands along the coast from Oslo to Bergen. Dutch, British, and Norwegian ships sailed into the fjords and loaded the finished planks directly from the sawmill, without having to pay for additional transport. In the mid-17th century, the growth of the sawmill industry was threatened by deforestation. The sawmills had to move inland. More important, the industry was regulated by the law of 1688, which allowed only a fixed number of sawmills the privilege of export and also set a production ceiling. Towards the end of the 18th century, sawmill technology was improved through the introduction of the Dutch fine-blade saw that cut four planks at once. It was more energy efficient and reduced the waste of raw material. Around 700 privileged export sawmills were in operation at the time.

Neither the fish export trade nor the sawmill industry required heavy investments. Norwegian merchants and farmers with some capital to invest were able to participate. The skills involved were mostly traditional and were widely diffused. The development of

The Sandaker sawmill in Oslo, ca. 1840. The sawmill industry spread rapidly along the southern coast of Norway during the 16th century. From a painting by Thomas Fearnley.

Source: The Norwegian Museum of Science and Technology.

Norwegian shipping followed a similar pattern. Merchants exporting fish and lumber often transported the goods on their own ships. The growth of the Norwegian merchant marine was boosted by the British Navigation Act of 1651. The legislation meant that the lumber export was reserved for Norwegian and British ships. Norwegian shipping also grew during wars and conflicts among the major maritime powers Holland, Britain, and France in the later 17th century. While the Norwegian fleet consisted of 240 ships in 1670, the number had quadrupled by 1696. In the first half of the 18th century, the size of the fleet remained stable, but grew rapidly after 1750 when the major powers were again involved in prolonged conflicts. Between 1750 and 1807, the fleet grew from 500 ships to 1,600. Shipping was strongly linked with shipbuilding, which experienced even faster growth than shipping by supplanting imported tonnage and exporting ships to Denmark.

The population of Sweden (including Finland) was almost four times larger than that of Norway in 1800. But the tonnage of the Norwegian merchant marine was one-third larger than the Swedish. By that time 12,000 Norwegians were employed in the international shipping trade, twice the number employed by all mining enterprises in the country.

It was in capital-intensive industries heavily dependent on advanced imported technology that the Danish Crown made its presence felt in Norway. During the age of mercantilism, the Danish king developed policies to foster trade and industry. In many respects the two countries constituted a common economic market. A division of labor evolved, based on each country's comparative advantages in natural resources. Denmark was well suited for agriculture, while Norway could harvest the sea and its forest and mineral resources. The Crown actively promoted industrial projects in Norway that would contribute to the overall balance of trade, to substitute for imports of products from other countries. Most of these projects had to do with mining and metal production.

Christian III was the first Danish king to invest in Norwegian mining on a large scale. His mother's brother, Count Johan Friedrich of Saxony, in 1537–1538 helped the king recruit the skilled Saxon mining engineer Hans Glaser to evaluate promising geological prospects in Norway. On the king's behalf copper mines were claimed in the county of Telemark; a couple of them were located in the wilderness far from the nearest fjord. The Saxon expert also helped frame the first Norwegian mining law according to Saxon tradition. The prospects for extracting precious metals from the Norwegian mountains seemed promising, and the king sent Glaser back to Saxony to recruit skilled miners, buy equipment, and solicit capital. However, private investors willing to put their money into a risky venture in the far north were hard to find. The king therefore had to raise the necessary capital himself.

Workers proficient in the arts of mining and metallurgy were easier to lure away, and in the summer of 1539, the first contingent of Saxon miners arrived in Norway, followed by a second the year after. Many of the foreign workers had a hard time adapting to the local environment. Some of them returned to their native country after a short stay, while the ones who remained with their families came to form foreign enclaves within the broader society, and the king recruited a German priest to serve their spiritual needs. Mining received wide-ranging privileges from the Crown, privileges that often clashed with local traditions and economic interests. Mining had to compete with the sawmills for wood resources, and farmers were more or less forced to transport wood and provisions for the new mining industry. Grain had to be imported from Denmark, and after arriving in Norway, had to be transported on horseback far inland. The ores proved less rich than expected, and operational results were disappointing. In 1549 the king tired of throwing good money after bad, and closed down the whole operation. Most of the skilled workers returned home, although some moved to other parts of Norway where the prospects seemed more promising.

The mining enterprise started by Christian III in the mid-16th century is the first major example of technology transfer on a grand scale to Norway. Persons having the necessary skills as a tacit component of their training were imported, and it was not part of their assignment to train local Norwegians. If more workers were needed, they too were imported. This pattern of technology transfer was used time and again by both the Crown and private investors throughout the mercantilist period.

When rich silver ore was discovered at Kongsberg in 1623, Christian IV claimed the mining rights. In the next few years, the king invested heavily to develop the discovery. The town of Kongsberg was built from scratch and given special mining privileges. New and improved mining and metallurgical methods were once again acquired by importing knowledgeable foreigners. The king operated the enterprise on his own until 1627, when most of the shares were sold to private interests. In 1683 the state took over again and remained the sole owner until the works were closed in 1805.

Silver production at Kongsberg was without doubt the most important mining enterprise in the country during this period, but copper mining also took place in eastern and central Norway, the enterprise at Røros being the largest. Ironworks were established in the southeast. Approximately 20 ironworks were in operation at one time or another in this period. The quality of Norwegian pig and bar iron compared well with Swedish iron, and most of the production was exported. Iron production peaked during the Napoleonic Wars, and 17 ironworks had 22 blast furnaces in operation in 1806–1807.

While the Crown was heavily involved in silver and copper mining, private investors dominated the iron industry. But all the metal-producing industries enjoyed royal privileges to make production economically feasible and to counter foreign competition. In the mid-18th century, the metal-producing industry employed 8,000 persons, half of them at Kongsberg. Management positions and the highest paid jobs generally were held by immigrants or descendants of foreign skilled workers.

The economic policy of promoting industries by granting them privileges, by handing out government subsidies, or by building high customs barriers for certain products was continued during the 18th century. The establishment of Norwegian glassworks and the methods by which they acquired technology further illustrate some of the general points we have made. In 1739 some Germans and Danes, the king himself among them, put together capital to establish the Norwegian Company (Det Norske Compani), with the explicit aim of supplying Norway and Denmark with glass products that had been imported and with strengthening the overall balance of trade.

The first glassworks, Nøstetangen, was erected in 1741, 60 kilometers west of Oslo, followed by six additional glassworks in the next decades. All of them were located near abundant forests for fuel. The partners were in charge of operations until 1776, when the state took over, an arrangement that lasted until 1824. In the first years of operation, the main product was bottles, but with time high-quality crystal was also produced. In a relatively short time, the new Norwegian industry could supply the domestic and Danish market with all kinds of glass, except for plate glass. To help bolster the growth of the industry, the government banned the import of glass to Denmark and Norway in 1760.

The glass industry was developed in a planned fashion and with the aim of supplying both markets with sufficient quantities of all types of glass. As long as the works were unable to produce window panes, that goal was still not reached. The first skilled glass workers were recruited from Germany, although French glass blowers were also employed by the company. The British, however, were regarded as the leading producers of "crown glass," and in 1754 the company sent a young Norwegian to Britain to obtain information and recruit workers familiar with the "crown-glass" process. The prevalent attitude among nation states during mercantilism was that one nation gained riches only at the expense of another. Industrial espionage became an important instrument of technology transfer. The young Norwegian toured major glassworks in Britain, and despite being secretive about his real mission, he was found out and arrested by the British authorities. He escaped, however, and managed to take the information he had gathered with him. He also succeeded in recruiting "crown glass" specialists. The first five arrived in Norway at about the time that he was arrested.

This case has much in common with the others we have mentioned from the early modern period. It demonstrates how intimately technology transfer was linked with the migration of skilled workers. This was obviously the crucial element in turning major government-backed enterprises into successes or failures. The cost of acquiring specific technological processes might at times have been considerable and was certainly not justifiable measured with a modern yardstick.

Technology During the First Industrialization

Even before the first industrialization in the 1840s, the Norwegian economy was astonishingly integrated into the international economic system. The export of fish products, lumber, and shipping all played major parts in international trade. In 1800 about 20 percent of GNP was exported. Norway was no enclave economy, managed by foreigners who channeled profits out of the country. The economy was still agrarian, though. Only nine percent of the population lived in towns in 1800, increasing to only 13 percent in 1850, despite a tremendous population increase from 900,000 to about 1.4 million in the same period.

The establishment of industrial cotton textile factories formed the vanguard of the Norwegian industrialization process, as in many other countries. Several manufacturers started operations in the 1840s—the Nydalen and Vøien spinneries in Oslo (1845) and Arne Fabrikker outside Bergen (1846). In her study of the Norwegian textile industry in the mid-19th century, Kristine Bruland has shown, using the Hjula Weavery (1856) in Oslo as her major case, that the textile manufacturers ordered complete "technological packages" from British textile engineering firms. The Norwegian entrepreneurs provided capital and had the necessary commercial and marketing skills, but lacked textile engineering know-how. This posed no major bottleneck, however, since a number of British textile engineering firms were more than willing to supply their Norwegian customers with a whole range of services. British firms channeled

The Vøien cotton mill, Oslo, during the 1850s. Establishment of cotton textile factories in the 1840s formed the vanguard of industrialization in Norway.
Source: The Norwegian Museum of Science and Technology.

state-of-the-art technical information to Norwegian entrepreneurs. They helped them choose the right equipment, built it, designed the production line, and sent over fitters and mechanics to set up the machinery in the factory. The machinery suppliers also helped the entrepreneurs recruit skilled British operatives and managers willing to stay in Norway for prolonged periods and to take on the responsibility of man-to-man transmission of their skills to the natives. This last point was perhaps the major difference with the forms of technology transfer mentioned earlier. Industrial entrepreneurs accepted all the help they could get from abroad to get started. All the same, they hoped to build up domestic know-how in the field to a level where they would be able to supplant the foreign technology in most respects.

The capital-goods industries were of crucial importance for the creation and diffusion of new technologies in the 19th century. The key to successful diffusion of new technology often lay in the ability of the capital-goods sector to assimilate and develop proficiency in the new machine technology and thus both to generate and adapt itself to the continually altering technological requirements of an industrializing economy. After 1850 Norwegian machine shops largely succeeded in this.

Nevertheless, Norway depended heavily on technology transfer from abroad. The British influence on the first machine shops was considerable. Between 1840 and 1860, a number of firms were founded in most of the larger cities—Akers Mek. Verksted (1841), Myrens Verksted (1848), Thunes Mek. Verksted (1849), Kværner Brug (1853),

Christiania Spigerverk (1853), Nylands Verksted (1854), all of them in Oslo; Nidelven Mek. Verksted (1843) in Trondheim; Bergens Mek. Verksted (1855) in Bergen; and Kristiansand Mek. Verksted (1855) in Kristiansand. Several of them imported machine tools, such as turning lathes, from Britain. Many of the firms that ventured into building steam engines and steamships, recruited British artisans and mechanics. After having accepted the position as chief engineer of Nidelven Mek. Verksted in 1843, John Trenery immediately returned to Britain to recruit skilled workers and equipment. Some of the founders and managers of the first shops had either worked at British engineering firms or studied British engineering practice for shorter or longer periods. The founder of Thunes Mek. Verksted, Halvor Thune, was sent to Glasgow to extend his technical knowledge. The future success of Christiania Spigerverk was largely based on the acquisition of the exclusive rights to use the nail-making machine patented by Ezra Jenks Coates.

During its first years of operation, most of the machine shops mentioned served the needs of the growing business community, delivering a variety of products. After some time, however, depending on engineering skills and market demand, they began to concentrate on the production of special types of machinery. The growth of flour milling, the introduction of new machinery in the sawmill industry, the upswing in construction building and shipbuilding all represented opportunities for Norwegian machine shops intent on substituting domestic products for foreign imports. Several of the general-purpose machine shops in cities along the coast began to specialize in shipbuilding, a field in which British influence was particularly strong and lasting.

Some engineering firms, however, made a name for themselves delivering machines that served the needs of the wood-processing industries, including the fast-growing paper industry. In 1850 Myrens Verksted built its first steam engine and its first water turbine. Nine years later the firm shipped its first steam-powered sawmill. Myrens Verksted designed excellent wood planers, and by 1890 the company had shipped more than 400 planers to customers in Norway, Sweden, Denmark, Finland, Russia, Germany, Austria, Switzerland, Holland, Belgium, France, and Spain.

Wood pulp grinding was a German invention transferred to Norway in the 1860s. The first crude wood pulp grinder was introduced at the Bentse Brug paper factory in Oslo in 1863, six years after the invention was first introduced in Sweden and three years after Finland. While six mechanical wood pulp mills were in operation in Norway in 1870, by 1879 the number had risen to 27, and in 1890 the country had 51 such plants. The number of workers in the paper and pulp industry grew from 316 in 1870 to 7,606 in 1900. To put this growth in perspective, suffice it to say that the number of workers employed in Norwegian industry in the same period grew from 34,597 to 79,635.

The fast growth of the paper and pulp industry created a new market for machinery. By the time the demand for wood grinders skyrocketed, the original patent rights on the mechanical wood grinder in Norway had expired. Engineering firms, such as Myrens Verksted, Christiania Maskinverksted, and Kværner Brug, were all successful in copying the invention and designing improved wood-grinding machinery as fast as their foreign competitors. All wood pulp mills in 1890 except two had equipment made by Norwegian firms. Deliveries to the wood pulp industry constituted the backbone of Kværner Brug between 1870 and 1890. Kværner delivered its first wood grinder with a capacity of 130 hp to the Granfoss mill in 1873. The same year the company delivered its first turbine, a 230 hp Jonval turbine to Hoffsfos Træsliberi, another wood pulp mill.

At the turn of the century, the number of wood pulp mills built by Myrens Verksted exceeded that of any other producer in Scandinavia. The company built its first wood grinder in 1865, and had delivered 490 grind-

Wood grinders delivered by Myrens mek. Versted and installed at the Funnefoss wood pulp mill. The rapid growth of the Norwegian pulp and paper industry created a new and important market for Norwegian machinery.

Source: The Norwegian Museum of Science and Technology

ers of different sizes by 1905. Of these 107 were hot grinders, an American improvement the company copied and introduced in Norway in 1895. Over 40 years wood grinders were delivered to 59 customers in Norway, 14 in Sweden, two in Finland, and one in Spain. Myrens Verksted and other Norwegian engineering firms clearly controlled the home market and part of the Swedish market. At the turn of the century, the average capacity of wood grinders in Norwegian mills was the largest in the world.

The wood pulp industry had power needs of a hitherto unknown scale. They could only be met by turbines. The growth of the industry therefore stimulated the design and construction of water turbines. Even though the first water turbines constructed in Norway around 1850 were modeled on British types, in the next decades, a new generation of mechanics and engineers, many of them educated in Germany and Switzerland, introduced Continental turbine designs, such as the Fourneyron, Girard, Schwamkrug, and Jonval. It was hardly a coincidence that the first Girard turbine, with a capacity of 55 hp and shipped to the customer in 1867, was constructed according to the Swiss "Zuppinger system." Antonius Jensen, the son of one of the founders of Myrens Verksted, had specialized in the design of turbines at the polytechnical school ETH in Zurich. The first Norwegian-built Jonval turbine was delivered by Christiania Maskinverksted in 1868. The manager of this shop, Chr. B. Mohn, was educated in Germany.

Between 1870 and 1890, Myrens Verksted shipped 267 turbines to customers, of which 46 percent were installed to drive wood grinders. However, if total capacity measured in hp was used as a yardstick, the turbines delivered to wood pulp mills represented 74 percent of the total. The single largest turbine, delivered to Skotfos Træsliberi in 1886, had a capacity of 370 hp. This record was not surpassed until 1895 when Myrens delivered three 600 hp turbines to the wood-grinding factory Hønefos Brug. It coincided with the introduction of hot grinding, resulting in a leap in the size of turbines delivered to such plants. The delivery pattern for turbines at Kværner Brug from 1873 to 1890 was similar. Of 96 turbines delivered, 42 percent were shipped to wood pulp mills. Closer analysis shows that orders for turbines were often included in contracts signed for complete machinery deliveries to wood pulp mills.

By the end of the 19th century, Norway had built up a technological infrastructure that made the country self-sufficient in many fields. The country became less dependent on importing foreign capital equipment and skilled personnel to operate the new industrial technologies. Norwegian machine shops increased their knowledge and technological capacity and successfully adapted and further developed foreign technologies of central importance in the Norwegian socioeconomic context. During the first decades of industrialization, Britain was the main source of new technologies and the country where most Norwegians went for shorter visits to keep up with developments in their particular technical fields. In the last quarter of the century, there was a significant movement in favor of the Continent, especially of Germany. We have already seen that Germany was a major source of technical knowledge in fields, such as wood pulping and water turbines. Towards the turn of the century, new technologies became increasingly science based, which strengthened the reorientation. Last but not least, an impressive number of Norwegians graduated from German engineering schools. After returning to Norway to practice their professions, they often became enthusiastic spokesmen for German education and technology.

Technical institutions were established to support the emerging Norwegian industry. The first patent law was passed in 1839 and was revised in 1885, extending the protected period from 10 to 15 years. Technical societies dedicated to the advancement of industry and technology began to flourish. Most influential was the Polyteknisk Forening, founded in 1852, which soon developed into the central meeting place for engineers and industrial entrepreneurs. In 1853 the society began publishing the journal *Polyteknisk Tidsskrift*, which became an important forum for the diffusion of new technologies. Towards the end of the century, *Teknisk Ukeblad* became the leading technical journal.

Following the traditions already established in Paris (1795) and Karlsruhe (1825), the first recommendation for establishing a polytechnical school in Norway was submitted to Parliament in 1833. The proposal was turned down with the argument that it would be too costly in relation to the small manufacturing sector it was to serve. The first technical school, training students in steam technology and engineering generally, was opened in 1855 at Horten. It had close ties to the naval yard there. This school and the three technical schools established in the 1870s, Trondhjems tekniske Læreanstalt (1870), Kristiania tekniske Elementærskole (1873), and Bergens tekniske Elementærskole (1875), gave the students a practical-oriented engineering education of three years. This was thought sufficient to serve the needs of Norwegian industry.

Many students, however, were not content with this lower-level education and went abroad to add one or two years of study to their Norwegian education. Most traveled to Germany to study at the Technische Hochschulen. Those interested in shipbuilding, however, often went to the Clyde area and the University of Glasgow. A smaller group, having a Norwegian gymnasium degree, went straight to Germany and spent four years at the Technische Hochschule.

Terms, such as engineer and civil engineer, were not widely used before 1850. In the following decades, however, much time and thought went into clarifying just what constituted engineering as a profession. At the same time, the number of persons calling themselves engineers grew. Norwegian students abroad were impressed by the many new and well-functioning engineering societies they came to know. In 1874 nine civil engineers educated in Germany decided to establish the "Society of Norwegian Engineers and Architects," an important step in the institutionalization of the engineering profession. In their outlook on the profession and the role of technology in society, the engineering association was heavily influenced by German ideals and practices. By 1916 the Norwegian Society for Civil Engineers had 1,735 registered members. We have biographical information on

TABLE 1
Norwegian Engineers Educated at Different German-speaking Technical Colleges

Polytechnicum Zürich	52
Techn. Hochschule Aachen	15
Techn. Hochschule Charlottenburg	99
Techn. Hochschule Danzig	4
Techn. Hochschule Darmstadt	28
Techn. Hochschule Dresden	104
Techn. Hochschule Hannover	60
Techn. Hochschule Karlsruhe	26
Techn. Hochschule München	30
Technicum Mittweida	17
Bergakademie Freiberg	8
Miscellaneous	15
Total	458

1,010 of them, and of these about 45 percent had studied for one year or more in Germany.

In the two last decades of the 19th century, an intensive debate on the question of higher engineering education raged in the Storting (Parliament), in newspapers, and in other forums. Despite public pressure from the engineering profession and students, the government did not wish to extend the level of technical education. In 1890 the Storting responded to public demands and allowed the three technical schools founded in the 1870s to extend education to four years. Ten years later the Storting took the final step and decided that Norges Tekniske Høgskole (NTH), closely following the German ideals of what a top engineering school should be like, was to be established at Trondheim. Classes started in the fall of 1910, and the first engineering graduates entered professional life in 1914–1915.

At the turn of the century, Norway had a much larger pool of well-educated engineers than either state or industry could absorb. In Norway, as in many other countries, it was not the new industry that employed engineers with the highest engineering education, but the government. Until the mid-19th century, military engineers were in charge of civil engineering projects: planning and building roads and bridges, improving canals, and building locks. The new foreign-educated civil engineers took over many of these responsibilities in the second half of the century. The growth of steam transportation also created new job opportunities. Scheduled steamship transport necessitated improving existing harbors and building new ones. To increase the safety of navigation at night, a chain of lighthouses had to be erected along the long coast. The first railway in Norway began operating in 1854. British engineers were in charge of its construction, and locomotives and all other rolling stock were imported from Britain. By 1875 the country had 550 km of railway lines, and Norwegian engineers employed by the state railways had taken over most responsibilities for construction and operation. This development continued, and Norwegian machine shops even began producing locomotives, at first using drawings bought from British builders. Between 1855 and 1870, the state built a network of telegraph lines spanning the country from north to south. Despite impressive industrial growth in the two last decades, the number of engineering positions available within industry was disappointing. Most firms preferred mechanics to engineers.

The supply of engineers was far in advance of demand. In the late 19th century, Norway actually exported many engineers to other countries. A typical career pattern for many included an engineering education in Norway followed by one or two years at a German engineering school. After graduation some were offered jobs with German firms, but most migrated to the United States or South America. Many stayed there for the rest of their lives, while others returned to Norway when the employment opportunities improved.

Electrification Permeates Norwegian Society

The new science-based industries were an integral part of the second industrial revolution in the 1890s. The introduction of technological innovations in electricity and chemistry was to trigger massive industrial growth in

Norway. Nationals and foreigners became aware that Norway, with its many large and undeveloped waterfalls, had a more promising potential for hydroelectric power production than any other country in Europe. Norway had always imported its coal, and the electrification of factories and railways would in itself be of great economic importance. But it was above all the country's potential as a producer of electrochemicals that caught the imagination of engineers and investors.

The first plans for huge power plants were conceived and put into practice in the mid 1890s. The first factories, Hafslund and Borregaard, situated on each side of the Sarpsfossen in southeastern Norway, were designed to produce calcium carbide. Operations started in 1899 and were soon followed by power stations and carbide factories at Meraker and Notodden. Market demand for calcium carbide grew fast in the early 20th century, but the number of carbide factories and their capacity grew even faster. Constantly falling prices led many factories to adapt their production equipment to related but more promising processes requiring vast amounts of electric energy. Production of ferro-alloys and other types of electrochemicals grew rapidly. the first aluminum works began operation between 1906 and 1909, and during the First World War, production expanded rapidly.

The development of one of these energy-intensive companies loomed larger than any other, and to many it came to stand as the symbol of the age of science-based industry in Norway. Norsk Hydro, incorporated in 1905, was to exploit a Norwegian invention, the Birkeland-Eyde process for the manufacture of synthetic nitrogen fertilizer. It was an electrothermic process requiring large amounts of energy. The nitrogen in air was converted to nitric oxide in an electric reverbatory furnace. Large power plants were built at Notodden and Rjukan, and by 1912 the unprecedented sum of 100 million kroner had been spent on power stations, factories, and houses for staff and workers. This was the largest industrial project ever undertaken in Norway, and in a few years, Norsk Hydro could pride itself on having become one of the world's largest producers of synthetic nitrogen.

Different uses of hydroelectric energy became a powerful motor in the Norwegian industrialization process for decades. From 1900 until the First World War, the large, private electrochemical and electrometallurgical companies were clearly the forerunners in this development. Products, such as calcium carbide, synthetic nitrogenous fertilizers, ferro-alloys, and aluminium, began to weigh heavily in the Norwegian export statistics after 1910. The large industrial projects to exploit Norwegian water power required investment capital on a scale that the domestic market was unable to supply. The new industry was largely financed by Swedish, French, German, and British capital. Even though the capital involved seemed breathtaking to many, the cost of developing the waterfalls in this period was exceptionally low, and most plants could claim to be among the cheapest hydroelectric power plants per hp in the world.

In the years before the First World War and during the war, an increasing number of towns and municipalities invested in power stations and transmission and distribution lines. Electric power production was evidently a field where economies of scale could be reaped. Towns and neighboring rural communities joined forces to build large power plants. The price of electricity decreased, and demand increased. During the war neutral Norway had problems importing coal and kerosene, leading to high prices. This in turn resulted in an unheard-of demand for electric energy, and there was strong political pressure to develop power plants serving the needs of the general population rather than the private electrochemical corporations. By the mid-1920s, the capacity of power plants serving the utilities surpassed the capacity of the electrochemical industry.

With some notable exceptions, such as the Birkeland-Eyde-process at Norsk Hydro, most factories relied heavily on foreign patents and processes. The electrochemical in-

TABLE 2
The Development of Norwegian Water Power (1,000 hp)

	Paper & Pulp Industry	Utilities	Electrochemical Industry	Other Types of Industry
1895	75			25
1900	90	8	10	36
1907	117	31	48	42
1913	180	125	390	60
1922	220	694	766	140
1928	203	1050	975	69

dustry with its growing power requirements challenged electrical manufacturers and turbine builders to design larger and more economic electric generators and turbines. In the decade after 1895, the construction of turbines changed drastically. With the coming of electric power, customers were no longer content with the customary efficiencies, but required the turbine manufacturer to guarantee an efficiency of 90 percent or more in conjunction with high specific speeds. Swiss and German companies perfected the design of the Francis and Pelton turbines and emerged as the leading manufacturers worldwide for a time. All the large contracts for turbines to the large power plants built in Norway between 1900 and 1910—Hafslund, Rjukan, Notodden, Tyssedal—went to foreign firms, such as J.M. Voith and Escher Wyss. The same was true for deliveries of generators and electrical equipment. The Norwegian market was dominated by firms, such as Siemens, AEG, ASEA, and Brown-Boveri. None of the smaller Norwegian manufacturers could muster the capital and R&D resources necessary to compete for the large contracts.

Nevertheless, the execution of the many large building projects created the career opportunities that the engineers had been dreaming of. The Norwegian engineer and entrepreneur Sam Eyde was involved in bringing together natural resources, technology, and capital in connection with several of the largest power projects, the establishment of corporations, such as Elektrokemisk and Norsk Hydro, being the most important. The companies he managed recruited top-level engineers in an unprecedented number—civil engineers, hydraulic engineers, electrical engineers, mechanical engineers, and chemists. Engineers living abroad got an opportunity to return home, while many already working for a Norwegian company or institution were lured away by what seemed to be greener pastures and often better wages. Talented engineers showed great skill and inventiveness in hydraulic engineering, and several made contributions to improving current electrochemical methods and processes. The first major power projects in the 1890s were focused mainly on harnessing the low waterfalls with masses of water in southeastern Norway. After the turn of the century, interest turned towards waterfalls with high heads in western Norway. Water often fell several hundred meters from the mountains to sea level in a fjord. Precipitation from large areas in the mountains was gathered in huge natural or artificial basins, where the potential energy was stored, to be used when needed. Norwegian engineers introduced new concepts in dam building and tunneling. But the crucial factor and the weakest point in connection with high-pressure power projects were the steel pipes leading the water from the storage basin to the turbine. Norwegians contributed to the safety of the

The Tyssedal power station in Hardanger began operation in 1908. It is a good example of a west coast power plant. The head of water at the top of the mountain to the fjord below is 465 meters. *Source:* The Norwegian Museum of Science and Technology.

system by inventing automatic closing devices in case of breakdowns.

By 1920 what might be called the electrical system permeated Norwegian society. Power plants, transmission lines, utility companies, manufacturing enterprises, and regulating state agencies constituted the core of the system. But it involved more than the hardware, the machines, and the processes associated with it. People and organizations also gave the system its momentum.

Large technological systems represent powerful vested interests. Electricians and electrical engineers certainly developed specialized skills integral to the system. But civil engineers, building contractors and their workers, and even machine shops in time came to depend on the further development and continuation of the electrical system.

Norwegian machine shops that built turbines illustrate the point. Domestic manufacturers neglected the new market for turbines to be installed in electric power stations and concentrated too long on the need for turbines in the wood pulp industry. Norwegian turbine builders lagged behind technologically and were unable to meet the specifications set by the power stations. The manufacturers tried hard to bridge the gap before 1914, but could not compete with the Germans and Swiss until the war years. Then both domestic and foreign firms had more orders than they could deal with. In addition, the traditional German suppliers of both turbines and electrotechnical equipment had trouble keeping the delivery schedules. Domestic suppliers reaped high profits, and some of it was plowed back into technological development to strengthen their relative competitive position. In the fall of 1920, the boom ended, and Norwegian industry went into a slump. The frenzied development of waterfalls came more or less to a standstill. Capital was hard to come by, and owners of power plants built during the war had a hard time servicing their debts. Still, the electrical system had gathered strong momentum. Turbine builders, contractors, electrical manufacturers, and engineers continued to press for further development of Norwegian water power. When banks all over the country went bankrupt, when electrical utilities had to get state subsidies and loans to stay operational, new converts to the gospel of electricity were hard to find. The electrical system in itself was intact; only the pace of its development was slowed down. Its protagonists had to accept the economic facts of life and wait for the business cycle to turn upwards again.

Incremental Technological Change in the Interwar Period

The interwar period was disappointing for technological enthusiasts in Norway. Rising production and profits during the First World War led many industrial firms to invest in new factories and product development. With the coming of the economic depression in the 1920s, demand fell dramatically, and many industrial companies could no longer service their debts. The industrial sector experienced bankruptcies and reorganizations. Industrial production stagnated in real terms and stayed below the 1916 level for most of the 1920s.

Industrial managers had a hard time just keeping their firms running, and there was certainly little room for investments in new technology. The two decades from 1900 to 1920 had been a golden age for engineers and had increased the prestige of the engineering profession. In the 1920s many engineers found themselves without jobs again or had to perform tasks that people with less education could have mastered just as well. "The Class of 1919" at NTH, the Norwegian engineering school, illustrates the point. More than 50 percent of the engineering graduates emigrated, and students began calling the civil engineering department the Norwegian-American Line, after the shipping line that took passengers and cargo between Norway and the United States.

From the beginning of the 1930s, however, the prospects for Norway's industry began to improve, and during the next two decades, the country experienced industrial growth on the same scale as early in the century. While industry contributed 30 percent of GNP in 1935, by 1950 it had risen to 40 percent. Homesick Norwegian engineers living abroad could find positions with Norwegian firms that enabled them to return home. This time, however, it was not necessarily the power-intensive export industry that needed their services, but smaller firms producing for the domestic market. The diffusion of technological innovations to small and medium-sized firms was of great importance. Incremental technological changes led to impressive industrial growth. Many craft-oriented furniture shops, for instance, developed into modern mass-producing furniture factories. Well-designed, standardized furniture produced on electrically powered special purpose machines conquered the market. A new awareness about the importance of rationalizing the production process took hold. Engineers in favor of Taylorism got their first opportunities to practice their knowledge on the factory floors.

Social Democrats and Engineers Join Forces to Build Industrial Norway

Despite the German occupation of Norway from 1940 to 1945, most industrial firms continued to grow until they were no longer able to secure raw materials. In the first postwar years, the major task was to reconstruct the industrial capacity and improve productivity.

The Tysland-Hole pig iron furnace at Svelgen, 1928. The develpment and adaptation of the Søderberg continuous self-baking electrode to pig iron production during the 1920s led to the resurrection of the Norwegian pig iron industry, using electricity instead of charcoal.

Source: The Norwegian Museum of Science and Technology.

During the 1950s Taylorism was at its height, strengthened by funding through the American Marshal Plan.

The electrical system got the revival the engineers had been waiting for. The government, led by a strong Social Democratic Party, regarded the further development of Norwegian water power as being of crucial importance to any plan for industrial growth. This time around the state was to play the leading role, not private industry dominated by foreign capital.

Since the opening decade of the 20th century, the debate on resurrecting the old pig iron industry by using electricity instead of charcoal had gone through several phases. During the First World War, the Norwegian company Elektrokemisk built up a strong R&D department in electrochemistry. One of its major projects was to develop a continuous self-baking electrode for furnaces producing ferro-alloys. The engineers succeeded, and in 1918 Elektrokemisk acquired patents on the Søderberg-electrode in several countries. The company continued to develop the electrode by adapting it to other types of electrometallurgical and electrochemical processes and succeeded in exploiting the patent rights on the world market. This invention, one of the major ones made by Norwegian engineers in this century, laid the foundation for the survival and further growth of Elektrokemisk in the interwar period. In 1922 Elektrokemisk formed a joint development-venture with Christiania Spigerverk to try to adapt the electrode to the production of pig iron, and in a few years, an economically feasible pig iron furnace had been constructed, known as the Tysland-Hole furnace.

During the depression years, any plans for using this technology as the core of a pig iron plant with sufficient capacity to supply the domestic market came to nothing. After the Second World War, the project was given high priority, and in 1947 Norsk Jernverk in Mo i Rana was established, with the state as the sole owner. The state also participated actively in the development of the large aluminum company Årdal & Sunndal Verk, with works at Årdal, where the Germans had started building during the war, and at Sunndal. The government generally favored projects that involved capital-intensive export industries.

In 1930, 2.8 million Norwegians consumed

7.630 GWh electricity, 60 percent of which was consumed by power-intensive industry, while the civilian sector consumed only 20 percent. By 1950 consumption had doubled. Civilian consumption, comprising private and public services, households, and agriculture, had increased to 33 percent, while power-intensive industry now accounted for 45 percent. Development of waterfalls continued at a great pace during the 1950s, and from 1950 to 1960, consumption rose from 14.817 GWh to 27.509 GWh, the power-intensive industry and the civilian sector growing at the same pace.

Large industrial projects led to many contracts for domestic firms producing capital goods, such as turbines, generators, and other types of electrotechnical equipment. Shipbuilding also got a strong revival after the Second World War. In the interwar period, Norway maintained its position as one of the world's largest shipping nations, and Norwegian shipowners pioneered in the technological change to modern diesel-powered oil tankers. Domestic shipyards, however, were unable to compete for new building contracts and lagged behind in the interwar period. Norwegian shipowners chose to build a large percentage of their new ships at Swedish shipyards. After the Second World War, the Swedish yards continued to excel. But at the same time, Norwegian shipyards invested in modern equipment and methods and eventually acquired a larger percentage of new building contracts from Norwegian shipowners.

On the whole the Norwegian government and private industry were able to keep pace with international technological developments that were seen as important in the Norwegian socioeconomic context. If a new technology was regarded as important by a firm or the country generally, some strategy was usually devised to acquire it. In the new fields of electronics, for instance, the government through publicly financed R&D institutions played a key role. By 1970 the per-capita income in Norway was among the highest in the world.

Offshore Oil Creates a New Technological System

Norway also seemed to have promising prospects for exploiting hydrocarbons from beneath the North Sea. In 1962 the Norwegian authorities were approached by foreign multinational oil companies concerning licences for offshore exploration and production of petroleum on the Norwegian continental shelf. The first permits were granted in August 1965, and Esso spudded the first wildcat well in the summer of 1966, using the American-constructed and operated semisubmersible drilling rig "Ocean Traveler." Norway had no experience whatsoever in oil exploration, and at first, all necessary skills and technologies were imported. The main agents in the technology transfer were the multinational oil companies holding the concessions or companies on contract to the oil companies. As long as the Norwegian authorities and private firms looked on exploration as a highly risky venture, probably resulting in no commercial finds, little was done to acquire skills associated with the offshore technology. The foreign companies with their exotic technology had come and would leave again, and Norwegian society would be no worse off than before.

This was not to be. In the fall of 1969, Phillips Petroleum Co., as operator for a group of companies, found the Ekofisk Field. News about the discovery began to spread in the spring of 1970, and after having drilled two appraisal wells, the company announced that the first giant North Sea oilfield had been found. Finding reserves capable of supporting large-scale production in Western Europe amazed Europeans and excited even the most cynical members of the industry. Gas had been found earlier, off southeast England, but this was the first time oil in large quantities had been found.

The Ekofisk discovery triggered frenzied exploration drilling both around Ekofisk and on geological structures further north, in considerably deeper waters, where the weather was even harsher than the operators had experienced so far. In a few years, a string of giant oil and gas fields were discovered on both the British and Norwegian continental shelves—Forties (GB 1970), Frigg (N 1971), Brent (GB 1971), Piper (GB 1973), Ninian (GB 1974), Statfjord (N 1974).

The Ekofisk discovery changed the course of Norwegian history. The planning, construction, and installation of permanent production facilities meant that the oil companies were there to stay. Considerable income would flow into government coffers through royalties and taxes. Exploration and production of petroleum would mean the establishment of a new offshore-related industry. The earlier policies were no longer viable. Norwegian society and industry wanted to participate fully in the development. Policies to bridge the technology gap had to be devised and executed.

The gross production value of oil and gas as a share of Norwegian GNP grew from 0.5 percent in 1974 to 15 percent in 1981. During the same period, exports of oil and gas as a share of total Norwegian exports grew from 0.8 percent to 31 percent.

We have earlier described how the electrical system came to influence Norwegian society in the early 20th century. During the 1970s offshore industry and technology came to form the dominant technological system, permeating the whole economy and thereby the whole society. This does not mean that offshore-related technologies excluded all others, far from it. My point is that offshore technology, or technologies that in some way or other were relevant to it, set the tone for technological development in Norway in the late 1970s and 1980s. Today, at the beginning of the 1990s, no alternative can challenge the supremacy of the offshore technological system.

No Norwegian universities and technical colleges had ever taught their students subjects related to the exploration and production of hydrocarbons. During the early 1970s, oil companies had to import foreign geophysicists, petroleum geologists, reservoir and petroleum engineers. Norwegian institutions of higher learning introduced oil-related courses, and by the end of the decade, the first graduates began entering professional life.

In 1972 the Storting established an Oil Directorate and a state-owned oil company, Statoil. The former was to control and supervise all petroleum activities on the Norwegian continental shelf, dealing with subjects, such as conservation and resource management, on the one hand, and safety, on the other. Statoil was to be the nation's major political instrument in safeguarding a fair percentage of the economic rent. When Shell discovered the Brent Field in British waters, the geological experts thought that the structure might extend over and into the Norwegian sector. The Norwegian government allotted Statoil 50 percent ownership in the blocks in question, with Mobil as the operator. Drilling in the spring of 1974 confirmed that there was no connection with the Brent Field. Instead a new giant oilfield was discovered, the Statfjord Field. Through the development of this discovery, Statoil got a head start, and in a few years, the company built up its organization and expertise to become a fully integrated oil company. Today Statoil is the largest corporation in the country.

Both the Oil Directorate and Statoil played crucial roles in acquiring foreign offshore technologies and adapting them to the Norwegian socioeconomic and environmental contexts. But many private companies also showed great skill in taking up new ventures and making original contributions of their own. In the early 1970s, Norwegian shipowners entered the offshore-drilling business. They ordered new semisubmersibles designed for North Sea conditions. Most of them went into joint-ventures with American drilling entrepreneurs to get the necessary drilling expertise. The Aker Group made headlines with their semisubmersible rig Aker H-3, designed and manufactured by the yard. The design was also licensed to foreign shipyards.

During the first years after the North Sea was opened up as a new and significant offshore oil and gas province, the oil companies adapted offshore oil technologies that had been developed in the Gulf of Mexico. Fixed production platforms there were always steel structures. A steel jacket was fabricated in a shipyard, floated out to the exact location in the sea, and then placed on the seabed. Following this process, the steel deck and different modules, such as the drilling rig itself, the living quarters, or the helicopter deck, were lifted onto the jacket by sea cranes.

The same methods and procedures worked well in the North Sea as long as new gas fields were found in shallow waters. Due to the harsh weather, however, with gales and waves almost constantly battering the steel platforms, improvements soon had to be made to strengthen the structures. Should the fixed installations on these new fields make use of the traditional technological solutions and use steel structures, or was new technology called for, considering the significant depths and the demanding environment? Norwegian engineers familiar with advanced structural concrete technology felt that concrete gravity platforms would be an ideal solution.

During the planning of the development of the Ekofisk Field, the need arose for a large storage tank. Normally tankers would be loading oil directly off the field, but during bad weather, oil would have to be stored for several days. The alternative was to discontinue production. Phillips chose to build a huge storage tank designed by a French company. Three Norwegian building contractors with considerable experience with concrete formed a joint-venture, Norwegian Contractors, to build the tank. Its foundations were laid in a drydock in Stavanger. After some progress had been made, the dock was flooded, and the structure towed out to deeper water. Continuous pouring of concrete was very important,

Towing out the base and platform legs prior to the addition of the steel deck for the Statfjord A Field, summer of 1976. The Condeep concrete gravity platforms came to symbolize the scale of oil production in the North Sea. *Source:* Statoil

and to keep a convenient work height, the tank was ballasted down in step with the building progress.

The concrete storage tank was a success, and the engineers at Norwegian Contractors began to play with the idea of designing and building concrete gravity platforms to replace the traditional steel structures. At the bottom it was to have a gravity base, to be placed on the seabed. The cells of the base could either be filled with water or be used as a storage area for oil. From the base three or four legs could be erected, and on top of them would come the steel platform with the production equipment. The building of the concrete structure could be done in one location, while steel deck and modules could be manufactured at different locations. When finished, the concrete structure could be towed to a deep fjord, be ballasted down to almost sea level, and be mated to the deck, resting on pontoons at a level only a little higher. Then water would be pumped out of the platform legs, and the platform with the deck on top would rise out of the sea. This design concept made it possible to complete a much greater part of the installation work inshore before being towed out to the offshore location. Besides being convenient, it would also be cheaper, labor costs inshore being only one-third of what they were offshore.

Would the oil companies be willing to consider the design seriously? Most experienced engineers employed by the operators were used to thinking in terms of steel solutions. Admittedly, the environmental conditions in the North Sea were far more demanding than anything hitherto experienced. European yards with any experience in doing offshore work were already booked for several years, and the traditional shipyards were busy building supertankers. There was a shortage of steel and rising steel prices. The operators in the North Sea were in a hurry to develop their resources. These and other factors contributed to a situation where traditional offshore engineers were more willing than previously to consider new technological solutions.

In the summer of 1973, Norwegian Contractors was able to land two contracts for concrete gravity platforms on the British sector, one for Shell at the Brent Field, and one for Mobil at the Beryl Field. In the late seventies, when the huge structures were towed out from the Norwegian coast to their respective locations by power tugs, pictures of them made the front pages and TV screens not only in Norway and Britain, but worldwide. Concrete gravity platforms became the symbol of oil exploration in the North Sea.

Throughout this article I have tried to concentrate on technology transfer and adaptation as the central theme. During the mercantilist period, technology transfer was closely linked to the migration of skilled foreign workers. Discussing the first industrialization and electrification era, I made it a point to show that the domestic technological capacity had increased to a level where it was seldom necessary to import any more foreign experts. The favored strategy was to send Norwegians abroad for a shorter or longer time to get acquainted with a new technology, bring the technology and knowledge back with them, and adapt it to the Norwegian environment. At several occasions during the last century, Norway actually had a larger pool of skilled engineers than industry and society at large needed. When oil exploration started in the 1960s, this was an activity hardly any Norwegian had thought of. At first Norway was wholly dependent on importing both skilled experts and equipment. The pattern of technology transfer frequently used in the mercantilist period was resurrected in connection with oil production. When it became evident that oil production was to be an important new industry, the strategies for technology transfer used time and again during the 20th century were put into action by both the state and private companies. Norway, with four million inhabitants, will always have to import most of the technologies needed. To keep up with the international development, the country has to be creative and flexible in executing the transfer process. In many respects the way oil technology was transferred in the first phase may be looked upon as outdated for an advanced industrial nation. But it proved effective. Norwegian contributions to the development of new technology have usually been made in fields closely related to its main natural resources. This is certainly also the case in the field of offshore technology.

Further Reading

Kristine Bruland, *British Technology & European Industrialization. The Norwegian textile industry in the mid-nineteenth century* (Cambridge, 1989).

Kristine Bruland (ed.), *Technology Transfer and Scandinavian Industrialization* (Oxford, 1991).

Fritz Hodne, *The Norwegian Economy, 1920–1980* (London, 1983).

Gunnar Nerheim, "Technology transfer from Germany to Norway at the Turn of the Century," in *Bürgertum und Bürokratie im 19. Jahrhundert. Technologie, Innovation, Technologietransfer*, Bericht über das 2. deutsch-norwegische Historikertreffen in Bonn, Mai 1987 (NAVF, Oslo 1987), pp. 159–174.

Technology in Sweden

Jan Hult

Technology for a Military Power, 1650–1720

One of the most dramatic developments in Swedish history was the country's emergence as a strong military power in the early part of the 17th century. A prelude to these events was the breakup in 1521 of the Kalmar treaty of 1389, by which Sweden, Denmark, and Norway had formed a political federation. Finland had been part of the Swedish realm since the 15th century. Under the forceful kings of the Vasa dynasty, Sweden gradually expanded its influence and territory, so that after the Thirty Years War in Germany, the Baltic Sea had become a Swedish lake. The country soon had to face a series of technological problems in order to maintain its status as one of the Great Powers of Europe.

The Great Copper Mine

The cost of the military and political expansion was partly carried by the yield from the Great Copper Mine at Falun, one of the world's then richest known copper deposits. Another early asset was the silver mine at Sala, which, however, was almost depleted towards the end of the 16th century.

New methods and new technology were constantly being tried at Falun. The traditional method of ore extraction was fire-setting. A pile of firewood was placed against the mine wall. The fire made the rock more brittle, so that it could be broken into pieces using sledgehammers and crowbars. Early attempts to use powder were not successful, but eventually powder came to be used in conjunction with fire-setting.

Water flow was a constant problem in the mine, and new, efficient drainage pumps, developed by the mine engineer Christoffer Klem, were important improvements. Likewise the old windlasses, operated by miners or by horses, were being replaced by new water-powered hoisting machinery.

Christopher Polhem, a leading figure in the Swedish history of technology, came to Falun as mine engineer in 1693. In the next 16 years he developed and improved a series of technical innovations, among them the rod engines,

Hoisting machinery at the Great Copper Mine designed by Christopher Polhem.

Engraving by Jan van Vianen, 1695 or 1696, in Sten Lindroth, *Gruvbrytning och kopparhantering vid Stora kopparberget Vol. I*, Uppsala 1955, facing p. 334.

by which the power from a water wheel could be transferred over a long distance. Several of these novelties required careful maintenance. After Polhem had left Falun in 1709, they tended to be neglected and were gradually taken out of operation. Much of the mining operations returned to traditional methods.

Polhem himself had moved on to create a works at Stjernsund in southern Dalecarlia, where he tried out various new ideas in manufacturing. Among them was a machine for the production of gear wheels for clocks. Clockmaking was then still a handicraft, and Polhem's industrially produced gear wheels were never really approved of by clockmakers. The machine itself has been preserved at the Tekniska Museet in Stockholm, where one may also see Polhem's "mechanical alphabet." This is a series of wooden models showing various simple kinematic elements. Polhem saw every machine as made up of combinations of such elements, just as a poem is basically a combination of letters and words.

Wallon Gunsmiths

An important immigration into Sweden of miners, ironworkers, and blacksmiths from Belgium took place in the 17th century. They established themselves in industrial communities in Uppland and Östergötland, such as Dannemora and Finspong. Being mostly Calvinists they did not easily assimilate into the Swedish Lutheran population, and thus tended to live in rather closed communities for several generations.

The skill they brought to Sweden was most important for improving both the technology and quality of iron production. The iron ore from the Dannemora mines was extremely pure, and its "Wallon iron" was much sought after. A large part of Dannemora iron was later exported to England for the production of crucible steel, notably at Sheffield.

Under Louis de Geer, one of the most successful Wallons in Sweden, the ironworks at Finspong became an important gun foundry. Later on, the Finspong works also developed the production of brass and other metal alloys, important in later stages of Swedish industrialization.

Naval Architecture

When Gustaf Vasa seized power in Sweden in 1521, the navy was essentially a mercenary organization. Ships and crews were made available to the state for a contracted period of time. This implied great savings, but it also meant that technical knowhow relating to the building and maintenance of naval ships was gathered in a few private hands. The naval architects looked on their knowledge as private property, which was often transferred within the family, from father to son.

The dangers of this system were obvious, and Gustaf Vasa ordered the creation of state-owned shipyards at several naval ports. When the king died in 1560, the military navy comprised 56 ships and the merchant navy around 60.

From the middle of the 17th century, British naval architects were invited to work in Sweden at the state-owned shipyards. A naval base was created at Karlskrona in 1680, and this also became the place where the first regular training of naval architects took place. Later on, in the 18th century, an institute for naval architecture was created at Karlskrona. Its era of greatness came in 1781 when Fredric Henric af Chapman was made director of the institute. He was born in Sweden of English parents but went to study naval architecture in England, France, and Holland before returning to Sweden. During his time at Karlskrona, naval architecture was given a scientific basis. His work *Architectura Navalis Mercatoria* (1768) is a world-famous classic of naval architecture with its detailed drawings of some 60 ships of different types and also with detailed figures relating to buoyancy, etc.

Chapman developed the theoretical means to study the stability, strength, and maneuver-

Work in a shipyard.

Source: Åke Claesson Rålamb, *Skeps Byggerij eller Adelig Öfnings Tionde Tom*, Stockholm 1691, in Gustaf Halldin (ed), *Svenskt skeppsbyggeri*, Malmö 1963, p. 97.

ability of ships as well the means to select suitable dimensions and materials for different types of ships. His book was later translated into French, English, and Russian.

Besides drawings, models of ships were also used as patterns in the education of naval architects. A great collection of such models was housed at Karlskrona. A similar collection of models of machinery, the *Royal Chamber of Models*, had been created in Stockholm, housed in the Royal Palace. It was open to the public for study, and its aim was to inspire new ideas and new applications. Christopher Polhem's mechanical alphabet was part of this collection. As has been remarked by Svante Lindqvist, "the models in the Royal Chamber of Models were prototypes of technology. They were images of technological aspirations in Sweden during the eighteenth century."

Palaces for the New Nobility

Swedish military officers sometimes returned from the war on the Continent with enormous booty. They were ennobled, and started to build palaces fitting their new rank in Swedish society. A period of active building gave rise to monuments still standing, stately homes for the new nobility. Palaces at Läckö, Skokloster, Tidö, and Ulriksdal are examples. The architecture was often inspired by Italian, Dutch, and French patterns.

An impressive building, "Makalös" (meaning matchless), was erected by the field commander Jacob De la Gardie facing the royal palace in Stockholm. This was the largest privately owned building in the capital, and it embodied the self-consciousness that characterized so many members of the new nobility.

Many new cities were also founded in the era when Sweden was a Great Power. In all 32 new cities were founded during the 17th century. Among these Gothenberg later became the second largest city in Sweden. Stockholm, the capital, expanded rapidly. From 1620 to 1660, its population rose from about 9,000 to 50,000 people. Its narrow, winding streets were gradually replaced by an orthogonal network of streets.

Sweden's era as a great military power ended abruptly during the reign of king Charles XII. Initial victories during his campaign in Russia did not prevent a disastrous defeat at Poltava in Ukraine. On 30 November 1718, the king was killed by a bullet outside the Norwegian castle of Fredrikssten.

Manufactories, Wood, and Water Power, 1720–1815

Sweden adjusted remarkably well to its new status as a former military power. Having lost most of its possessions on the other side of the Baltic, the country now consisted essentially of Sweden proper and Finland.

The mercantile system was gradually being adopted in a Sweden turning to a more peaceful existence. In Britain the first wave of an industrial revolution was beginning to transform society, and several Swedish travellers brought home new ideas. Among them were Jonas Alströmer and Mårten Triewald.

A Failure at Alingsås

Jonas Alströmer, fond of adventure, had set out for England as a young man. He had some previous experience as a bookkeeper in a business house, and he soon managed to carve out an existence in London working with an export company selling woolen textiles to Sweden. The idea of creating a large-scale textile industry in Sweden soon occurred to him.

He then set out on a "study tour" in Britain and on the Continent that was really industrial espionage. In Holland he managed to buy some advanced textile machinery, notably, two stocking-stitching machines. Export of such equipment was strictly forbidden, but Alströmer dismembered the machines and had them sent to Sweden hidden in the hold of a Swedish merchant ship. The destination was Gothenberg, and from there the equipment was transported to the town of Alingsås, some 40 miles inland.

Alströmer's plan was to create here, in his home town, a large-scale textile manufactory. He was successful in gaining strong economic support from the faction then in power in the Swedish Parliament. Parliament also granted exemption from military service for young men in Alingsås and its surrounding countryside, so that there would be no shortage of manpower in the Alingås works. Production started in 1724, and at its peak the works employed over 1,000 men and women.

Except for the imported Dutch machines, most of the equipment at Alingsås was traditional spinning-wheels and hand looms. The novelty in this enterprise was its magnitude. The Alingsås Manufakturverk became one of the largest industrial workplaces in the country.

The plan was to sell the products at markets in the surrounding regions and also in Stockholm. But the quality of the cloth coming out of the Alingsås works turned out to be inferior to much homemade cloth, and sales were slow. Economic difficulties were soon threatening the bold enterprise, but Parliament came to the rescue, and production continued. But more and more of the products accumulated unsold in the great warehouse in Alingsås, and when a new faction came to power in Parliament in 1765, support was suspended, and the good days were over.

Work then continued for many years but with a steadily decreasing production until 1824, when one of the earliest and boldest Swedish attempts at large-scale industrialization finally ended.

Jonas Alströmer himself spent most of this time in Stockholm lobbying Parliament but also mingling with various members of high society and with scientists, among them Carolus Linnæus and Mårten Triewald. He apparently preferred this company to struggling with the chores of being a textile mill director at Alingsås.

A Failure at Dannemora

Mårten Triewald had also set out for England as a young man. After a period of unemployment he became the secretary of a businessman and managed to remain in London. There he attended public lectures on experimental philosophy held by the curator of the Royal Society, John T. Desaguliers. Triewald himself was later elected a member of that academy. When his secretarial appointment ended, Triewald was named superintendent of one of the first Newcomen steam engines at Newcastle-upon-Tyne. He thus acquired first-hand knowledge of this new technology and got the idea of erecting a steam engine in Sweden, where drainage of mines was a common problem. In 1726 he returned to Sweden and within a few months had secured financial support for this enterprise from the mineowners in Dannemora.

The mines at Dannemora were on the edge of a lake, and a constant flow of water gave increasing difficulties as the excavation proceeded to ever-deeper levels. Windlasses and horse whims tended to be insufficient. A pumping engine of the Newcomen type gave great promise. The erection of the machine started in 1726, but difficulties soon presented themselves. Sweden did not possess the skill and capacity to build a Newcomen steam engine without considerable delay. When the machine was finally installed, other problems caused difficulties.

Its fuel consumption was very great, and quantities of firewood and twigs had to be delivered. The pumps would often freeze on cold days. The entire system was too complicated for the unexperienced people at Dannemora to manage. But Mårten Triewald himself, who really knew how to run the machinery, was often absent from Dannemora. He preferred to be in Stockholm with his good friends Jonas Alströmer, Carolus Linnæus, and a few others, all of them interested in Natural Philosophy. The steam engine adventure at Dannemora ended on a sour note; Triewald was sued by the mine owners who had suffered a great economic loss.

There are clear similarities between the industrial failures at Alingsås and Dannemora. Both were attempts to establish a modern enterprise of a kind that had proved to be successful in England, and both failed. Sweden was not yet prepared to adopt and to

Mårten Triewald's steam engine at the Dannemora iron mine.

Source: Engraving by Eric Geringius 1734, in Svante Lindqvist, *Technology on Trial,* Uppsala 1984, page 284.

assimilate a technical system of this magnitude and complexity.

But Alströmer and Triewald were more successful in another venture, the one that had drawn them to Stockholm.

The Royal Academy of Sciences

In 1739 Alströmer, Triewald, and Linnæus came together with two members of the House of Nobility in Stockholm to form what is now the Royal Swedish Academy of Sciences. The pattern was the Royal Society in London, and Triewald gave public lectures at the House of Nobility modelled on those he had attended in London.

The lectures were on experimental philosophy and included demonstrations, but the ambitions of the academy soon became very utilitarian. Investigations were performed, and studies were made, on ways to utilize various energy resources in Sweden. At this time there was a widespread fear that forests in Sweden would be depleted due to increasing demand from ironworks and other industrial enterprises.

Experiments with different types of water wheels were made, and the results were published in the *Proceedings of the Royal Swedish Academy of Sciences.* These were all written in Swedish, the aim being to bring the findings to the attention of the Swedish people. Different types of ploughs were described, and their qualities were compared to the benefit of farmers. It is not likely, however, that these results were put into practice. There was not much contact between these learned men in Stockholm and the population of Sweden, which was overwhelmingly rural.

Machine Tools, Steam Engines, and Railroads, 1815–1870

In the wake of the Napoleonic Wars, Sweden lost Finland in 1809. In 1815 she left her very last possessions on the continent, and the boundaries of the country have since remained unaltered. In 1814 a cease fire had been declared in the campaign in Norway, the last act of war that has been undertaken by the Swedish army.

Industrialization was still proceeding slowly in Sweden. Lack of capital for investment was a contributing factor, even though the Swedish East India Company had been very successful in creating such assets. The negative aspects of the rapid British industrialization were also well known in Sweden, and many leading personalities were issuing warnings.

A comprehensive land reform, starting in 1803, had led to an increase in arable land by some 15 percent. At the same time, a rapid increase in population set in, from 2.4 million in 1815 to 4.1 million in 1865. But job opportunities did not increase at the same rate. Large-scale emigration to the United States of America acted as a safety valve. Sweden was spared mass unemployment.

Göta Canal

The loss of Finland was a great shock to the Swedish people. It served as a warning that no more victories should be expected on the battlefield. The future lay in peaceful endeavors. Improved means of transportation within the country would increase contact between people and would help emerging industries.

The idea of a canal across the country was not an new one, but it took a decided man like Admiral Baltzar von Platen to carry it through Parliament and to get the digging started. He invited Thomas Telford, the famous Scottish canal builder, to survey the landscape and propose a layout. In a few months, Telford had made a complete outline, and the project was under way.

Swedish army units working on the Göta Canal.

Source: From a painting by A.C. Wetterling, the National Museum, Stockholm, in Eric de Maré, *Swedish Cross Cut*, Malmö 1964, p. 63.

The entire operation was organized in military style. Several regiments of the Swedish army exchanged their rifles for spades and wheel barrows. It took 23 years to complete the canal, and 60,000 men in all were engaged in the project, the largest in Swedish history.

The canal had military, productive, and symbolic importance. It made it possible to move naval vessels from the Baltic Sea or the west coasts to a safe haven inland. For this purpose a large fortification was built at Karlsborg on the shore of Lake Vättern. Rooms were also prepared for the Royal Family, the government and Parliament to assemble there in case of war. This "Central Defense Doctrine" was one of the driving forces behind building the canal.

Its importance for industrial production was obvious. Raw materials could be transported to industries on the canal, and finished products could later be transported to customers. The pattern was England, where a network of canals had been created that facilitated the transport of heavy industrial products.

Solar energy driven hot air engine due to John Ericsson.

Source: John Ericsson, *Contributions to the Centennial Exhibition*, New York 1876, facing p. 53.

The symbolic importance of the Göta Canal was to rally the nation around a remarkable project, something to be proud of, and that would make it possible to forget the tragic loss of Finland. As one poet said, the aim should be reconquer Finland *within* Sweden. The projected cost of the canal was exceeded several times over, but in 1832 the "nation's ditch" was finally completed.

Among the men working on the canal were the Ericsson brothers John and Nils (who spelled his family name Ericson). John later left Sweden for England and America, where he acquired fame as a prolific inventor. Nils ended up being responsible for the Swedish state railroad program, which had been instituted by Parliament in 1853. The railroads, ironically, soon threatened the canal. They permitted more rapid transportation, and they could be extended in various directions, forming a network that would eventually cover much of the country. Göta Canal never attained the importance that its creators had expected. Today the canal is a great asset to tourism in Sweden, but industrial raw materials and finished goods travel on rails or roads.

Motala Werkstad

Even though much of the toil at the canal consisted of manual digging, there were other tasks as well. Spades and crowbars had to be made, and bridges and sluice gates had to be designed and built. The initial plans were to use local iron works and foundries and to rely upon materials like fir and oak. After some years the canal company decided to speed up the work by mechanization, e.g., by acquiring cranes and a steam engine, that was used for running a drainage pump. The need for a larger foundry soon became obvious, but the directors were reluctant to invest more money.

The canal passed through several shallow lakes, which made it necessary to dredge their bottoms. Von Platen knew about the very effective steam power dredger that had been used by Telford on the Caledonian canal in Scotland, and he decided to order one for Göta Canal. But he also asked the manufacturer, the Bryan Donkin Company, to send one of its best workers along with the dredger. Thus it happened that Daniel Fraser arrived in Sweden in 1822. A workshop was arranged at Motala for service and maintenance. This was the beginning of Motala Werkstad, one of the most important mechanical industries in Sweden in the 19th century. Daniel Fraser soon became the head of this enterprise.

The plant expanded gradually, starting with a main workshop, a smithy, a carpentry, and a machine shop. Later on, a foundry, a plate works, and a template shop were added, as well as an assembly shop. Modern machine tools were installed, and the power source was a large water wheel.

During its first five years, all the products coming out of Motala Werkstad were equipment for the canal, but eventually more and more was produced for the market. Steam ships soon became an important specialty.

The advanced technology applied in some parts of Motala Werkstad called for skilled craftsmen. Recruiting soon became a problem because the plant was situated in a rural area with no industrial traditions. The solution was for the company to organize a trade school within the works, where its own workmen were taught to use the new machinery. The Iron Founders Association in Stockholm also sent students there on fellowships to learn about the new workshop technology. These teaching activities were decisive for the future development of Swedish industry. Motala Werkstad became a hotbed for the emerging industrialization of the country.

Several future engineers and industrial leaders had been pupils at Motala in their youth. Among them may be noted: Carl Bolinder, founder (with his brother Jean) of Bolinders Mekaniska Verkstad in Stockholm, one of the capital's largest industries; Otto Carlsund, successor to Daniel Fraser as head of Motala Werkstad; Gottfried Kockum,

The Bolinder engineering works in Stockholm.

Source: Woodcut after a watercolor by C.S. Hallbäck in *Ny Illustrerad Tidskrift*, in Torsten Gårdlund, *Bolinders. En svensk verkstad*, Stockholm 1945, p. 77.

leader of Kockums Mekaniska Verkstad in Malmö, which also became an important shipyard; Alexander Lagerman, prolific inventor of machinery for production of safety matches; and Wilhelm Tham, leader of Husqvarna Vapenfabrik, an important armory.

Until 1850 the market value of the entire production at Motala was larger than that of the rest of the Swedish mechanical industry. One decisive reason for this success was its modern organization. The work was led by three supervisors, two of whom were British. They were paid by the month and were relatively independent. The traditional system with masters, journeymen, and apprentices was never applied. Every employee was paid by the hour according to diligence and skill.

Göta Canal has sometimes been called Sweden's oldest and longest technical university. Many young men were involved in the project, in the field work, or at Motala. They all acquired knowledge about civil or mechanical engineering, about the organization of work teams, etc. When, later, they brought this knowledge back to their home districts, it was made useful, albeit on a much smaller scale.

Dynamite

In the early 19th century, excavation in Swedish mines was mostly made by powder blasting instead of fire-setting. Blasting increased productivity and improved ventilation in the mines. But drilling boreholes was a tedious job, being entirely manual. The drill rod had to be sharpened 40 or 50 times for each meter drilled. Humidity in the mines was a problem; the powder had to be kept absolutely dry.

The discovery in 1846 of the great blasting power of fluid nitroglycerine raised hope for new developments in mining technology. But problems soon appeared. The fluid was very sensitive to shocks and could easily explode with careless handling. After a few years, all sale, transport, and use of nitroglycerine in Sweden were forbidden.

Alfred Nobel, a chemical engineer, had been experimenting for years on ways to harness the unpredictable fluid, when in 1866, he came upon a solution that would eventually make him one of the richest men in Europe. By soaking the nitroglycerine into a kind of organic sand called "kieselguhr," he obtained a doughlike substance that could be handled without danger, but that still possessed an

Early 20th-century open pit mining at Malmberget, close to the Artic Circle.

Source: Norrbottens Museum, Luleå.

enormous blasting power. He called it *Dynamit*. Earlier he had invented a detonator that secured remote handling of the blasting.

The timing of Nobel's invention was perfect. The new steel processes, developed by Henry Bessemer in 1856 and Pierre and Emile Martin in 1863, had led to an upsurge in steel production, and steel works were asking for more iron ore. Iron mines had to increase their production accordingly, and dynamite helped them do it. Abundant availability of steel at low prices encouraged railroad construction, which, again, led to demand for dynamite for cuttings and tunnels.

In his testament of 1896, Alfred Nobel set up the five prizes that every year place Sweden in the limelight.

Schooling

An important law was passed by Parliament in 1841 and put into effect in the year after. Schooling was made compulsory for all Swedish children. Every parish had to establish a school. Teachers had to be trained at special teachers colleges. Even though it took time for the reform to be implemented in all parts of the country, the decision of 1841 was a most important one.

Various types of schools had existed long before this, notably schools preparing for studies at one of the two universities in Uppsala and Lund. But very few students attended these schools. From 1842 the situation was completely changed. The aim was to give every Swedish child the ability to read and write and to do elementary mathematics.

The market for books and newspapers increased accordingly, causing an upsurge in the paper making and printing industries. The importance to the emerging industry of this development was obvious. Young people would now come to industrial jobs able to read instructions and drawings and write notes and do simple calculations, all things indispensable in an industrial environment. It is generally agreed that the Swedish school system was one of the reasons for the smooth transformation of Sweden from an agrarian to an industrial nation.

Already in 1825 a state committee had been appointed to study means of improving higher education in Sweden. The permanent secretary of the Academy of Sciences, Jacob Berzelius, proposed that courses in natural science topics should be compulsory in all higher education in Sweden, but he found scant support. He was more successful in giving a scientific touch to the engineering schools that had been founded in Sweden earlier in the century.

A mining school at Falun had opened in 1819. In Gothenburg an engineering school had started in 1829 as prescribed in a will by William Chalmers of the East India Company. Both these schools had high scientific ambitions, due to the efforts of Berzelius. By contrast the Technological Institute in Stockholm moved away from theoretical studies, which eventually led to the expulsion of its director in 1845. The institute was then reorganized and soon developed into a modern high-level institute for engineering education. In 1876 Parliament decided to rename it "Kungliga Tekniska Högskolan" (KTH), "the Royal Technical University," to conform with the technical universities ("Technische Hochschulen") in Germany.

Towards the end of the century, so-called "folk high schools" were started in several places. These were voluntary organizations, modelled on the ideas of the Danish philosopher Grundtvig, offering courses above those given in the compulsory school system. They were one of the first exponents of the populist national movements that were later to play an important part in forming Sweden as we know it today. Among these should be mentioned various nonconformist churches, the temperance movement, the labor movement, and the cooperative movement.

Birth of an Industrial Nation, 1870–1914

For a long time, Sweden seemed reluctant to step into the industrial era. Established cultural personalities, such as the author and historian

Erik Gustaf Geijer and archbishop Johan Olov Wallin, had warned against what had occurred in England with its slums and misery.

But the foundations of industry were being laid all the same. The guild system had been abolished in 1864, and new laws guaranteeing freedom of trade had been instituted. Investment capital could be borrowed from new banks. Railroads were being built and interconnected into a network.

The national government had also been reformed. The four Estates of Parliament had been replaced in 1866 by a Lower and an Upper House, a first step towards the democratic breakthrough that would follow early in the next century.

Around 1870 the country was ready for an industrial transformation.

Electric Power

The ability to transport energy from one place to another is one of the main considerations in planning for industry. In the early days, industrial plants had to be located near waterfalls in rivers, where water power was available. The Swedish term for waterfall is *fors*; hence names like Bofors, Söderfors, or Forsbacka for places with old steel and ironworks. Likewise, iron works, *hammers*, sometimes gave names to industrial sites like Hallstahammar, Morgårdshammar, or Hammarby.

The advent of the steam engine changed this dependence upon water power. Industries could now be located in more suitable places. Factories were equipped with a steam engine that was used to drive transmission shafts in the ceiling. Pulley belts distributed power to machine tools, spinning machines, or looms.

The next step in the development of energy transportation was based on the discovery of electric induction by Michael Faraday. It was now possible to transform mechanical energy into electrical energy. Generators driven by water turbines or by steam engines could supply electric current to electric lamps or to electric motors. Electromechanical industries began to develop, such as Allmänna Svenska Elektriska Aktiebolaget, (ASEA), which was founded in 1883. Under the name Asea Brown Boveri, ABB, it is now Sweden's largest industrial company.

All forms of power transmission suffer from losses, be it a rod engine system or an electric power line. According to the laws of physics, the losses in an electric transmission are lower the higher the voltage, and therefore alternating current has become important. It can easily be transformed to high voltage and then be retransformed at the receiving end of the power line. An early disadvantage of using alternating instead of direct current was the problem of electric motors. The solution to this problem was the three-phase system, invented simultaneously by Jonas Wenström at ASEA and by others. The first successful power transmission in Sweden with three-phase current was made in 1893 over a distance of 14 km.

The three-phase system immediately made a number of large water falls in remote areas in Sweden available for power generation to industries in other parts of the country.

Telegraphs and Telephones

An optical telegraph system, developed by Abraham Niclas Edelcrantz, had been important during the 1808–1809 war with Russia. Messages could be quickly sent from the outer Stockholm archipelago to the capital. But with the discovery of electromagnetism by Hans Christian Ørsted in 1820, the electrical telegraph developed quickly.

The first use of the telegraph was along railroad lines. Messages could be sent to the next station giving notice of the approaching train. The first telegraph connection between two Swedish cities (Stockholm and Uppsala) was opened in 1853. Only one year later, a connection had been established between Stockholm and Gothenburg (ca. 500 km). An underwater cable connected Sweden to the Continent from 1854.

The Swedish telegraph system was operated by the government. Hence the operators were state employees. In 1860 a woman, Elfrida Andrée, applied for a job as a telegraph

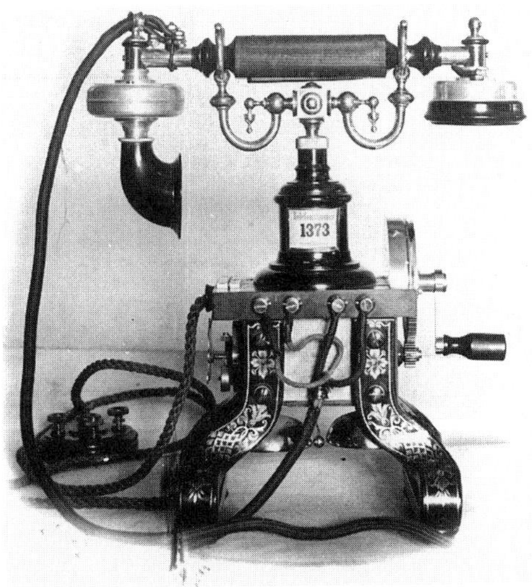

Lars Magnus Ericsson's desk telephone, 1892.

Source: The National Museum of Science and Technology, Stockholm

operator. She had passed the required examination showing proficiency in handling messages in Morse code, but she was not admitted. After four years of negotiations, her case was discussed in Parliament. A law was finally passed by which women were admitted as civil servants. Within a few years, most personnel at Swedish telegraph stations were women.

Lars Magnus Ericsson had a small repair shop for telegraphs in Stockholm. In 1876, the very year in which Alexander Graham Bell and Elisha Gray patented the telephone, he started to develop telephone equipment. Success was immediate, and he started the company, which under the name Ericsson, is now one of the best known in Sweden. By 1885 Stockholm had the highest telephone density in the world, both in relation to the population and in absolute figures.

Safety Matches

The invention of the safety match by Gustaf Erik Pasch gave rise to a world-wide industry that nearly became an economic monopoly. Its predecessors were wooden sticks dipped in a mixture of sulphur and yellow phosphorous, but they were poisonous and ignited too easily to be safe. The safety match had none of these drawbacks, and the factory that was started in Jönköping in 1845 soon became very profitable.

Initially the production in Jönköping was completely manual, but Alexander Lagerman designed and built a series of automatic or semi-automatic production machines. The first of these, which automatically filled 20,000 matchboxes in one day, was completed by 1877.

While designing machines for the match industry, Alexander Lagerman also designed type setting equipment as well as this automatic platen printing press.

Source: The National Museum of Science and Technology, Stockholm.

The culmination of this development was the 1892 machine system that produced 40,000 filled matchboxes in ten hours, starting with aspen logs, and needing to be supervised by only one or two people. From 1870 to 1900 the output of the Swedish match industry more than tripled, and the expansion continued into the next century. The annual export of matches at the turn of the century was almost 20 tons. Competition was now stiffening, and a Swedish Match Trust was formed that lasted until the bubble burst in 1931.

Ball Bearings

The bicycle found its present form in the 1880s, when the high wheeler gave way to the safety bicycle. Ball bearings were important new elements in this machine, and the rapid spread of the bicycle increased the manufacture of ball bearings. They soon began to find use in other kinds of machinery, such as transmission shafts in textile mills.

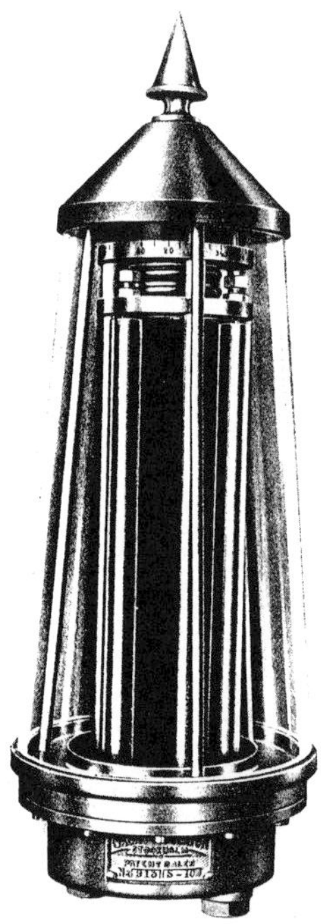

The Sun Valve, Gustaf Dalén's most admired invention. It automatically closes and extinguishes the acetyléne flame at dawn and relights it again at dusk.

Source: Erik Wästberg, *Gustaf Dalén,* Stockholm 1938, p. 114.

A large textile mill in Gothenburg, resting on a clay foundation, suffered from occasional settlings. The transmission shafts were bent, which damaged the ball bearings. This caused Sven Wingquist, chief engineer at the plant, to improve the standard ball bearing, permitting a certain tilting of the shaft within the bearing. The "spherical" ball bearing became the starting point of a new company, Svenska Kullagerfabriken (SKF), another cornerstone in Swedish industry.

The new ball bearings found widespread uses, but they could not be used in railroad cars. The load was too heavy, and the thumps from the rail joints soon destroyed the bearings. SKF then set up a research department to develop an improved type of bearing. After three years a spherical roller bearing had been produced that became the mainstay of SKF for many years to come. The development of this new bearing is one of the first examples in Sweden of organized industrial research leading to a new product.

Automatic Lighthouses

Acetyléne was the main light source in lighthouses and lightships at the end of the 19th century. It gave a much stronger glare than kerosene, which had been used in earlier times. A major problem, however, was the frequent need to refill the gas.

Two inventions by Gustaf Dalén resulted in major improvements here. The first was a valve that opened and closed automatically, thus producing light flashes only at intervals. This valve cut down the gas consumption by more than 90 percent. The other invention, called the "sun valve," made the lighthouse work only during the night hours, thus saving half the amount of gas again. Dalén was employed by the company AB Gasaccumulator (AGA), whose lighthouse equipment was sold all over the world.

Dalén's inventions made him very popular in Sweden. He was praised as an "engineering genius." In 1912 he was awarded the Nobel

Prize in physics much to the dismay of international science groups, who did not see any novel scientific ideas in his inventions.

Gage Blocks

One of the most important ideas in production technology in the 19th century was the concept of interchangeable parts. An early example was the production of small arms, such as pistols and rifles. Their assembly required the time-consuming fitting and adjustment of the various parts. When new machine tools, such as milling and grinding machines, had come into use, much higher precision could be achieved, and various parts could be assembled without prior manual adjustment. "The American system of manufacturing" became a common expression to designate this new workshop philosophy, which spread quickly to other branches of industry. Bicycles, sewing machines, agricultural machinery, etc. were all produced according to such standards.

The main point was to ensure that all parts that were meant to fit together fulfilled certain predetermined precision requirements. Accurate methods for the control of the dimensions of the finished products were needed. Such gages were often made in the form of carefully produced steel blocks that were used as references in the workshop control rooms.

Carl Edvard Johansson, working at Carl Gustafs Stads armory in Eskilstuna, had to use a number of different such gage blocks in his work as an inspector. When the Swedish army introduced a new rifle, many new gage blocks would have to be made, one for each particular dimension to be controlled. Every future

The de Laval cream separator *Alfa Baby,* launched in 1867, could skim 50 liters of milk per hour. After improvements in the late 1890s, the capacity was increased to 250 liters per hour.

Source: The National Museum of Science and Technology, Stockholm.

change would require new sets of blocks. He then got the idea of producing a set of standard blocks that could be combined to any desired size.

Johansson ground and polished his new gage blocks so well that they stuck together when combined. This implied that no accuracy would be lost in the process. The size of the built-up gage was equal to the sum of the separate gage sizes. The new gage blocks were first used within the armory, but were soon being produced for sale on the market. A company, Aktiebolaget C.E. Johansson, was formed in 1911, and export sales were brisk from the start.

The large-scale production of automobiles that had started in Detroit, USA, made good use of the new gage blocks. An American subsidiary company was founded in 1919, and Johansson himself was employed for some time by Henry Ford. His gage blocks are still called Jo blocks by workmen in many English-speaking countries.

The ball bearings, the lighthouse equipment, and the gage blocks were quite spectacular industrial successes coming out of Sweden at the turn of the last century. They were all examples of new products whose fabrication required extreme precision. Using present-day vocabulary they may be characterized as "hi-tech" products.

The Technological Fabric, 1914-1970

World Wars I and II, which devastated much of Europe, left Sweden unscarred. This promoted the strong development of industry and welfare, even though wartime economy put a strain on society. An almost exponential increase in gross national product (GNP) during 1900–1960 shows a doubling every 22 years. Whereas the population of Sweden increased from about 5 to 8 million in this period, the GNP increased from 2.5 billion crowns to 22.5 billion.

While many European countries experienced social upheaval, Sweden took the road of social understanding. Binding agreements secured peace and stability on the labor market. A "Swedish model" gradually took form, which came to be denoted as "the middle way."

Technology in Sweden developed "large-scale" features, characteristic of all industrialized countries in the 20th century. Mass production and the creation of large technological systems were key concepts in forming present-day Sweden.

A number of new industries had been formed towards the end of the previous century, based on inventions made by a few gifted individuals. Some of these inventors had an engineering education, others did not. As technological systems tended to become more complicated, more resources had to be directed toward higher engineering education and technical research, if Sweden wanted to stay in the forefront as an industrial nation.

Mass Production

The exploitation of the deep forests in northern Sweden began on a large scale when steam engines became commercially available. Log-driving on rivers brought timber to the mills, which were usually located on the coast, suitable for shipping. Production reached nearly twelve million cubic meters in 1970. Fiber board and plywood were new products with an expanding market in housing, in Sweden as well as abroad. Export volume in 1970 reached nearly 800,000 tons.

The breakthrough of the chemical pulp and paper industries occurred toward the end of the 19th century. New processes (sulphite and sulphate) had been perfected for the production of pulp from wood splinter, and large-scale mills were erected, mainly on the coast in northern Sweden. The total production in 1970 reached 8.5 million tons of pulp.

The demand for paper had increased steadily, one reason being the rising level of education caused by compulsory schooling, which was now generally the case throughout Europe. Pulp and paper soon became large export products. Newsprint became a main Swedish export. In 1970 three mills produced a total of 450,000 tons.

A gigantic vein of iron ore in Lappland had been of little value because of its high phosphorous content. The Thomas-Gilchrist steel process developed in 1875 had changed that, and large-scale exploitation had started toward the end of the 19th century. The Kirunavaara mine is located north of the Arctic Circle, and it was decided to build a railroad to Narvik, the ice-free port in Norway. Export of iron ore from Kiruna became an important source of income for Sweden throughout the 20th century.

Three large steel works were erected at Luleå, Domnarvet, and Oxelösund. Most smaller steel works in central Sweden, many of them dating from the 17th century, were found to be uneconomical and were shut down. Large-scale mass production had all but wiped out the traditional iron and steel culture before the end of the 20th century.

The American system of manufacturing was gradually and successfully adopted by Swedish industries. Also Frederick W. Taylor's principles of scientific management were planted in Swedish soil. Despite criticism from an ever-more politically conscious labor movement, these ideas were implemented in production and management. A well-organized Swedish industry was ready to meet the enormous demand of a Europe recovering from two devastating world wars.

National Power Grid

At the turn of the century, water power had already superseded steam power as the producer of electrical energy. The first exploitation of water falls for electrical power stations took place in southern Sweden. When a large power station was being projected in Göta älv at Trollhättan, the magnitude of the project was such that the government had to be engaged. Since that time most large power stations have been erected under government auspices.

A power station at Porjus in the Lappland wilderness was built between 1910 and 1915. Its main purpose was to deliver electrical energy to the mine fields in Lappland, but it eventually became an important link in the power grid that was soon beginning to take form. The ever-expanding pulp and paper industries in northern Sweden had increasing demands for energy, and new hydraulic power stations were erected in the 1920s. To minimize losses, the voltage on the power lines was gradually increased. The first 400 kilovolt power line was erected in 1952 from the new power station at Harsprånget.

More and more local power stations were then interconnected, and a main grid took form. In 1970 the total length of 400 kilovolt power lines in this network exceeded 10,000 kilometers.

Such long distance transmission is based on three-phase alternating current. Difficulties in transferring alternating high-voltage electricity in cables led Uno Lamm at ASEA to experiment with high-voltage direct current. An efficient system for underwater transmission of electric energy by high-voltage direct current in cables was the result. This system has found wide use, the first being a cable from the mainland to the island of Gotland in the Baltic Sea. Later on, high-voltage cables were laid between Sweden and Denmark, between England and France, and between islands in Japan.

Electric energy is now abundantly available to virtually all Swedes. The per-capita amount of electric power increased from 30 watts in 1914 to 940 watts in 1970. This corresponds to 30 light bulbs shining night and day for every Swede.

These data relate to electric energy. Taking all kinds of energy into account, the per-capita amount has risen from 220 watts in 1914 to 5,200 watts in 1970. The figures may serve to indicate the rise of the average living standard in Sweden during this period.

The first nuclear power station in Sweden was erected in 1963 at Ågesta south of Stockholm. This 80 megawatt unit, based on natural uranium and heavy water, was meant to be a pilot station, preparing for larger units. Even though the Ågesta station ran efficiently for ten years, it was then closed down, and the future development of the Swedish nuclear power program took a more conventional direction.

National Communications Grid

Whereas the telegraph system was originally operated under government auspices, private telephone companies were at first quite common. Only after both systems had been jointly placed under the government in 1920 could a systematic rationalization be achieved.

The telephone had originally been looked upon as a means of direct communication between two distinct places, such as an office and a workshop. Then came the idea of connecting several telephones by means of a switchboard. The invention of the telephone thus led to the technical system that is now a communications grid covering the entire country.

The first telephone station with an automatic switchboard, developed by L.M. Ericsson, was opened in Stockholm in 1924. Until 1950 all long distance telephone calls were manually connected by operators, but then area code numbers were introduced. In 1971 telephone traffic in Sweden was so extensive that every woman in Sweden between ages 18 and 55 would have been needed to work in switchboards if the automatic system had not been developed.

Direct dialling for international calls was also gradually introduced, and the cost of such long distance calls was simultaneously decreased. Sweden continued to be a leader in telephone technology. In the 1980s the national communications grid carried not only spoken messages but also messages between computers.

National Railroad Grid

The Göta Canal project never fulfilled its promises. The canal did not achieve the importance of which von Platen had dreamt. Railroads were the new means of transportation, and they soon overshadowed the canals. By World War I Swedish railroads were the dominant carrier of both goods and passengers.

The total length of rail tracks culminated in 1938 at 16,886 kilometers. The corresponding length per capita, about 2.5 kilometers, was surpassed only by Australia, New Zealand, Canada, Argentina, and USA. In 1970 the total track length had decreased to 12,669 kilometers, i.e., by about 25 percent. Private railroads, common in the early phases, were eventually taken over by the Swedish State Railroads. This transfer was essentially completed by 1960.

Electrification of Swedish railroads started early after small-scale tests in 1905. Electric locomotives hauling iron ore from the Lappland mine fields were found to be superior to steam engines. At the end of World War II, nearly two-thirds of the entire railroad grid was electrified.

Highways and Bicycles

The success of rail transportation was so great that many old highways began to fall into disrepair. The future belonged to the railroads. But a new means of highway transportation stopped the degradation of the highway system: the bicycle. It had found its now classical form towards the end of the nineteenth century, and in the year 1900, over 10,000 bicycles were produced in Sweden. Ten years later the production was more than 35,000, and the bicycle became a common sight on the highways.

The bicycle had a great impact on life in Sweden, in particular in sparsely populated rural areas. The temperance movement, the nonconformist churches, the labor movement, and other populist movements benefited strongly, now that people could travel some distance to meetings. The bicycle became a symbol of freedom, and the isolation of remote areas began to disappear. The increased mobility also gave more opportunities for jobs.

Payment by installments and mail order purchase were two novelties that helped to spread the bicycle to young and old throughout the country. As the number of bicyclists continued to increase, cries for better roads and highways were heard. In this way the bicycle literally prepared the way for the automobile.

Automobilism

At the turn of the century, there were ten or 15 automobiles in Sweden. An exhibition in Stockholm in 1902 spurred interest in this novelty, and attempts were soon made by various workshops to produce passenger cars and trucks. Early manufacturers were VABIS in Södertälje and Scania in Malmö. Other pioneer manufacturers were Tidaholms bruk and Thulinverken in Landskrona, but large-scale production of passenger cars did not commence until 1927, the starting year of the Volvo Company in Gothenburg.

The recession in the early 1930s hurt the company, but it still managed to widen its program to include trucks, tractors, and buses. The turning point came after World War II, when a new, small family car became a great market success. The success of the Volvo company encouraged the SAAB company, a manufacturer of military aircraft, to start production of passenger cars in 1950.

Volvo and SAAB developed side by side, and both companies managed to find good export markets. From 1945 to 1965, the number of passenger cars in Sweden (domestic as well as imported) increased from about 50,000 to 1.8 million, i.e., by a factor of 36. The car density in Sweden surpassed that of all other European countries.

With an ever-increasing number of cars, the old highway system was becoming inadequate. In 1953 the first dual super highway was opened between Malmö and Lund in southern Sweden. The standard of the old highway system was gradually updated with dual super highways, new bridges, etc. The changeover from left-hand to right-hand traffic in 1967 was performed with no increase in the accident rate.

By 1970 the limitations of the automobile were becoming evident to more and more people. The choking of inner cities and air and noise pollution were alarming signs. Automobilism as we know it cannot expand as before. But changes cannot come quickly. The sheer magnitude of the system indicates this. We find it everywhere in society: oil refineries, gas stations, highways, parking houses, car dealers, repair workshops, driving schools, insurance companies. . . .

Technical Research and Higher Engineering Education

An important event in the aftermath of World War I was the creation of the Royal Swedish Academy of Engineering Sciences (IVA) in 1919, the first of its kind in the world. Its aim was to promote research in various engineering fields, in particular relating to energy supply to Swedish industry. The academy itself eventually ran a few research laboratories, but only on a limited scale. Rather it became an intellectual base for deliberations over the technological future in Sweden. Qualified engineers and scientists were appointed as technical attachés at various Swedish embassies abroad. Their task was to send information back to Sweden about scientific and technological trends and breakthroughs.

In 1942 a technical research council was organized by the government to support research at universities and in industry. Later on a natural science research council and a medical science research council were set up. These councils had a very great influence on pure and applied research in Sweden.

Until 1937 the Royal Institute of Technology (KTH) in Stockholm was the only technical university in Sweden. In that year Chalmers Institute of Technology (CTH) in Gothenburg was also granted university status. A third technical university was created in Lund in 1961, to be followed by Linköping in 1969 and Luleå in 1971. In addition academic engineering education was instituted on a small scale in Uppsala.

In 1927 KTH had obtained the right to confer doctoral degrees, a decision at first not well received by the old universities in Sweden. CTH obtained the same status in 1937, and since 1971 an engineering doctorate can be obtained in six Swedish universities and institutes of technology. Research has become an integral part of all higher engineering schools in Sweden.

A Computerized Country

In 1947 the central tax authorities in Sweden had developed a personal code number system to simplify the handling of tax return forms. Every citizen in Sweden was to be identified by a ten digit number, the first six digits giving the birth date and the remaining six identifying the individual. The project was presented in Parliament, but there was virtually no debate about the desirability of such a system. It seemed quite harmless, and Parliament gave the project its blessing. The personal code numbers soon began to be used not only by the tax authorities but also by police, insurance companies, banks, driving license authorities, etc.

In that same year 1947, the transistor effect was discovered in the USA, and the microelectronic revolution got under way. Before long the modern computer hit the market, and Sweden was quick to adopt this new technology in administration and business, in the police, insurance companies, banks, driving license authorities, etc. All this has opened up possibilities for different authorities to connect various files and so to correlate various kind of behavior. The personal integrity of Swedes is at stake.

If the code number system had been proposed in Parliament today, it would probably not have been accepted for general use. The computerization in federal and local offices has been so thorough that a "full picture" of any citizen can be obtained without difficulty by referring to various computer files at different authorities. In 1973 a new agency, Datainspektionen, was created charged with the duty of protecting the integrity of Swedish citizens in an ever more computerized nation.

The personal code number system is now so ingrained in the fabric of Swedish life that it cannot easily be abandoned. Its advantages to administration and business are obvious, and very few people are openly expressing objections about it. In the wrong hands, however, it could be misused to an extent that cannot yet be foreseen.

Further Reading

Marquis V. Childs. *Sweden the Middle Way.* New York, 1936.

Tore Frängsmyr (ed.). *Science in Sweden. The Royal Swedish Academy of Sciences 1739–1989.* Science History Publications, USA, a division of Watson Publishing International. Canton, MA, 1989.

Carl G. Gustavson. *The Small Giant. Sweden Enters the Industrial Era.* Ohio University Press. Athens, OH, 1986.

Michael Lindgren. *Glory and Failure. The Difference Engines of Johann Müller, Charles Babbage and Georg and Edvard Scheutz.* MIT Press. Cambridge, MA, 1990.

Svante Lindqvist. *Technology on Trial. The Introduction of Steam Power into Sweden, 1715–1736.* Uppsala Studies in History of Science 1. Uppsala, 1984.

Eric de Maré. *Swedish Cross Cut. A Book on the Göta Canal.* Malmö, 1964.

PART TWO

The Industrial Heritage

Conservation and Preservation

The Industrial Heritage—Changing Perspectives

Marie Nisser

The conservation of industrial sites and buildings is a fairly new subject, but one that has expanded its frontiers enormously in both the geographical and scientific and museological senses as well as in conservation terms.

Activities on a larger scale began in the mid-1950s and may be connected with a growth in national pride in past industrial achievements. In Britain, the interest in industrial archaeology—as it is called—was growing in a new climate of environmental and historical awareness. Industrial archaeology was understood to be primarily concerned with the material remains of yesterday's manufacturing and transport that survived more or less intact on its original site.

During the 1960s there was a growing awareness in a number of countries of the importance of the subject. However, this appreciation was still only shared among a few enthusiasts. It took some time to find a platform for the activities connected with the subject.

The first international meeting on the conservation of industrial archaeology took place at Ironbridge in Britain in 1973. This may be taken as an indication that the subject had become important in a wider circle of European and other countries.

But industrial archaeology had still as yet found no secure foothold, either in Europe or in other parts of the world. Examples of conservation of significant sites for industrial and technological history could certainly be presented; but successful attempts to rescue them were outnumbered by failures to do so.

Public opinion, official conservation policy, organizations giving grants, as well as industries themselves had not yet begun to consider their industrial heritage as an important part of their cultural tradition. There were many reasons for this, but a decisive factor must have been the general development of industry and community. This must briefly be explained. For simplicity, let me confine myself to the Western industrial nations.

A Period of Industrial Reconstruction and Development

Even if it is a sweeping generalization, one can maintain that the decades after the Second World War were a period of hitherto unknown reconstruction, development, and expansion. For many branches of the economy, the climate was favorable. Reality was, of course, more complicated. Cyclical variations occurred even during the 1950s and 60s. Some branches of industry experienced considerable structural problems. But on the whole exporting industries, such as mining, steel, shipbuilding, pulp and paper, and chemicals developed satisfactorily.

Production was concentrated in plants of ever-increasing size. This continual structural reorganization was part of a process of rationalization that had been apparent on several occasions during the century. The advantages of large-scale production could be won by increasing plant size. Administration, marketing, and research were coordinated when companies merged. Capital was available for new investment; plants were modernized,and new machinery installed.

During this period a more advanced form of mechanization was introduced—automatic control of the production process. It was primarily progress in electronics and computers that made this possible. At the same time, innovations were introduced in other technical fields: the development of nuclear power, the petrochemical production of basic materials, the use of artificial fibers and plastics, printing by the offset method, to quote a few examples.

It is often maintained that technical developments during the years after the Second World War were more rapid and radical than during previous periods. This led to an extensive program of plant expansion and modernization and the closing down of smaller and unprofitable production units. Thus we lost many working environments and machines that otherwise could have reminded us of an earlier and important phase in our industrialization.

At the same time, most countries experienced a wave of new construction such as had never before been seen. In many cases it was the wounds of the Second World War that were rapidly healing; in other cases it was changes that were demanded not only in industry but in the whole community—the redevelopment of central business districts and a dramatic expansion of traffic systems and residential areas.

The Hostility Towards Industrial Conservation

Even if the economic trend was already levelling out towards the end of the 60s, the general tone of debate into the early 70s was one of optimistic expectancy. The idea of preserving old factories and workers' housing was often held up to ridicule—it was development and progress that were interesting. Moreover, there were many who had personal memories of unemployment and of hard years in inadequate factory buildings and unhealthy workers' tenements, of miserable towns, and of landscapes in destruction. The best that could happen would be to forget them. What could be more appropriate than to demolish the physical expression of these memories: the abandoned factories and housing from the early days of industrialism?

Industrial archaeologists in those days often confronted skepticism and a lack of understanding. In the early 70s it was still often difficult to break through the hostile attitude towards industrial conservation. But national policy towards the preservation of industrial monuments has varied a great deal. In the Scandinavian countries, there are examples of successful preservation schemes within industry throughout the 20th century, perhaps on a larger scale in Sweden than elsewhere.

The former German Democratic Republic and Czechoslovakia had early shown high appreciation of industrial monuments, as symbols of the history of the working class. Industrial heritage conservation received appropriate financial support in those countries; while on the other hand—unlike in Western Europe—castles, churches, and noble houses of the rich were played down. In some countries in southern Europe, such as Italy and Spain, industrial buildings and sites had long had relatively low prestige compared with the classical and artistic antiquities that are, of course, considered to be a nations' great glory.

The Changing Scene

The years after the 1973 Ironbridge conference were eventful both in general industrial development and more specifically for international cooperation of the industrial heritage.

Industrial development in the West did not follow the expected course. No one imagined at the end of the 60s that economic growth in the original industrial nations would come to a halt. A continued rise in the standard of living and an increasing industrial production were confidently forecast for Europe and other countries. Just the opposite happened. Optimism and the expansive years were replaced by economic crises and an uneasiness about future developments. Plants closed down in rapid succession. Even in economically sound branches of industry, structural readjustments have been comprehensive with large-scale unemployment the result. This economic crisis has been longer and deeper than any we have previously seen in Western industrial history.

During the 70s and 80s, not only the industrial sector but also community life in general have undergone radical changes. One effect is similar to the years of expansion, namely the closing down of factories. But there is also another effect concerned with the general attitude towards our past. Life at work and at home has been influenced, as well as our ideas and patterns of action. The period of transition is hardly over; much remains unclear and difficult to understand. Perhaps our new attitude towards the future is more resigned than it was at the end of the 60s. Our attitude to our historical heritage has also changed. In many situations the perspective towards the past has begun to be considered more important.

Our Identity with the Past

Paradoxically an increased interest in the industrial heritage has arisen in the wake of the industrial crises. Perhaps this is not so strange; in insecure times we tend to look back to the past. When we want to establish our identity in the present, we try to understand our connection with what has gone before us. A certain nostalgia is also there; it is not the years of crises, redundancies and closings down that have constituted the industrial historian's foremost field of work, but the technical achievements and periods of expansion and increased productivity.

In the same way, it was not our abandoned industrial areas from the last decades nor our slum districts that originally were industrial archaeologists' most sought-after study objects; rather it was the stately and undeniably important monuments from the early years of industrialism—mills, blast-furnaces, steam engines and other early technological masterpieces

But during the 70s we have seen the industrial archaeologist's domain change and the tasks be reformulated. First, the concept of industrial archaeology had to be accepted; it was later replaced in international contexts by the term industrial heritage. Thus it was emphasized that the subject had been broadened. In the infancy of industrial archaeology, the dominating interest was the history of technology in its narrow meaning. The field of interest then widened to include industrial environments. By the mid-70s there was discussion of preserving and documenting the actual history of working life in a wider sense. At about the same time, the tide of new building began to ebb, and the flood of demolition receded. Adaptive reuse and recycling became interesting.

Redundant factories found a new function and industrial areas a new activity.

New Tasks for Industrial Heritage Conservation

Industrial conservation's sphere of activity was thus extended; it has also come to encompass a longer period of time. In the beginning it was historical contexts that were studied and early industrial environments that attracted attention. Today contemporary documentation has come to be regarded as an increasingly important task. It is perhaps even more important for us to document the most recent changes; they will probably disappear more quickly than the historical monuments that still stand. There is, moreover, the ambition to achieve livelier forms of conservation and to find new forms of museum exhibits where knowledge of older skills can be kept alive. Here then can be new opportunities for employment, and people who feel alienated from the working environments of high-technology society can find satisfaction in new knowledge of old skills.

Despite these wider ambitions and new tasks, the study of tangible monuments and artifacts has continued to be our primary occupation. It must remain so, with industrial landscapes and their factories, residential areas, transport items, photos, documents, and so on. If we are going to preserve anything from the different phases of industrial development, it must be physical objects, but they must be explained and exposed in their social and economic contexts.

It could be claimed that our industrial heritage is the most eloquent witness we have of the history of man at work. A wide selection of industrial buildings, sites and landscapes must be preserved: not only those relating to the early history of industrial development but also environments relating to later phases. These will help us to understand our own time and explain how we reached it.

High ambitions on conservation have, however, not always been realized, and today we must admit that we can hardly hope to achieve such a program. The economic resources to manage a broad selection of colossal industrial plants preserved as monuments of modern industrial development are lacking. Necessity forces us to be extremely selective. Every country can preserve a number of mills, blast-furnaces, factories, and power-stations. The preservation of larger industrial units in a wider sense and in an international perspective must be considered. This is why the International Committee for the Conservation of the Industrial Heritage (TICCIH) has started to discuss industrial landmarks of international significance.

At this international level, it can hardly be hoped to solve such problems as legislation, restoration techniques, or internal policy questions; the conditions in the various countries are too particular. But we can probably make an important contribution by discussing which environments and plants are to be considered indispensable, not only nationally but also from an international point of view. Which are in fact the international landmarks of the industrial history of the world?

The Nordic countries have started to collaborate in this field, and hope that it will be possible to list and protect a number of industrial sites and buildings as a matter of common interest in our Nordic industrial heritage.

The Growing Importance of Documentation and Research

Seen in this context, the documenter's task becomes even more important. Our documentation will in most cases be all that is left of objects doomed to disappear. This demands a high standard of thoroughness, clarity, and versatility. As yet we have hardly developed satisfactory methods of documentation. Moreover, each branch has its particular problem; the documenter needs a comprehensive grasp of the whole as well as a knowledge of the details. The museum and conservation professionals cannot be expected to manage the tasks by themselves. In the beginning, documentation and research must be coordinated.

Research workers must cooperate if as many aspects as possible of the industrial heritage are to be documented. The interdisciplinary contributions to such combined research projects are important but insufficient.The teamwork between these professionals and the many enthusiastic amateurs who have so significantly contributed to a better understanding of the industrial heritage is is no longer enough. People from all levels of industry must also participate in documenting their own sector and its development. Those who work in an industry also have the necessary detailed knowledge.

The Future for Our Industrial Past

Many countries have now become involved in safeguarding and studying the industrial heritage. And new countries will be more and more involved in the future. The ideological backgrounds have differed between individual industrial archaeologists and historians and between their various countries. Motives for choosing the subject have varied, as have the objectives, approaches, contents, and methods. This has, however, been beneficial. It has broadened our activities, increased our knowledge, and stimulated a critical attitude, in the best meaning of the word. Seven international conferences have treated the subject; seminars, meetings, and conferences have dealt with special themes; national representatives have met to explore problems of conservation, documentation, and research; many books and journals have been and are continuously being published. All this has furthered appreciation of the industrial heritage.

In this sense we have grounds for being pleased with the results. Further activities have a secure foothold. But the conservation of industrial buildings, sites, and landscapes will continue to be a complex concern. And the commitments will be expanding along with the continuous transformation of the world.

Hirtshals
Fjerritslev
Ålborg
Nykøbing
Hobro
Randers
Viborg
Tange
Århus
Herning
Odder
Horsens
Hjedding
Esbjerg
Vejle
Odense
Egeskov
Stenstrup
Korsør
Egernsund
Helsingør
Nivaa
København
Roskilde
Køge
Næstved

Denmark

Ole Hyldtoft

Background

There is more to Denmark than just Hans Christian Andersen and Hamlet's Elsinore, and a good deal more to Danish industry than just beer, butter, and bacon. The earliest traces of industrial activity are flint mines from the Neolithic, while extraction of local bog ore was a vital power resource in the Iron Age. Shipping has ancient routes in Denmark, which is an island kingdom situated midway between Central Europe and the Scandinavian peninsula, and in between the North Sea and the Baltic Sea. The oldest known ships date back to the Mesolithic. Our heritage from the legendary Viking era includes five ships, among them two rare coastal merchant vessels, found in Roskilde Fjord and now on display in the Viking Ship Museum in Roskilde, which is a center for the study of early naval architecture and maritime archaeology.

In the Middle Ages, small windmills and water mills for grinding corn sprang up all over the country. The first really large industrial complexes were prompted by military considerations, as a means of strengthening the emergent nation state. Copenhagen's naval dockyard was founded in about 1550, and was soon followed by ancillary industries in the form of textile manufacturing, powder mills, and small arms factories in the city and its surroundings. Still extant from this period are the gable of the dockyard's anchor forge (1562), reused in 1619 for the building of Holmen Church, and Christian IV's stock exchange building in Dutch-Renaissance style (1619–1640)—another manifestation of the economic policy of the time. A new, mercantilist upswing followed at the end of the 18th century, when a large textile industry was established in the capital.

Modern, English-style industrialization set in about 1840. New products and new industries, such as engineering, brickworks, printing, garment manufacturing, and gasworks, grew up, while the historic textile industry faded into the background. At the same time, steam power and the machines connected with it became more widely used. The 1890s ush-

A 17th-century example of the recycling of industrial buildings. The anchor forge of the Copenhagen naval dockyard was built in 1562, to designs by Peter de Duncker. In 1619 the forge gable was reused for the chancel of Holmen Church. The forge faced the royal palace, hence its lavish decoration.

Photo: Ole Hyldtoft, 1991.

ered in a new wave of great innovations, this time mainly from the USA and Germany. Electric power stations, electric motors, and electric machinery symbolized the new upsurge, but the impact was far wider, from cement, diesel engines, motor vessels, motor cars, bicycles, bacon factories, margarine and shoe factories to cotton-spinning mills. A long line of new industries brought new methods of production, such as refrigerators in the food and drinks industries, composing machines in the printing industry, and high-speed steel, milling machines, and automatic welding in the engineering industry.

Thus the main outlines of Denmark's industrialization conformed to the general Western European pattern. The food and drinks industry was relatively strong, but development in Denmark was above all characterized by its breadth, without any particular sector dominating the others. The country had few industrial raw materials, and until the 1950s, Danish factories produced mostly for the home market. The greater part of industry, consequently, grew up in the old towns, which, like sea ports, occupied nodal positions in the transport network and which already had a concentration of market, labor, capital, and know-how. Something like half of Denmark's industry was located in Copenhagen. Next came the biggest of the provincial towns—Odense, Århus, and Ålborg.

One or two exceptions to this general rule should be mentioned. Most of the old water- and wind-powered factories were in the countryside. Nearly all dairies and bacon factories were established near the suppliers, with one or two dairies in every parish and a bacon factory in every hundred (i.e., district). The many knitwear factories around Herning in Jutland evolved from a flourishing, proto-industrial cottage production. Production of building materials, finally, was closely connected with local deposits. This applies to the quarries of Bornholm, the Portland cement factories of North Jutland (writing chalk), and the huddle of brickworks north of Copenhagen, at Steenstrup on the island of Fyn and at Egernsund in South Jutland.

Wind and Water Mills

From the 17th century onwards, a number of old water mills were harnessed for more innovative forms of industrial production, such as textiles, iron manufacturing, gunpowder, and paper. The importance of wind and water power was accentuated in about 1800 and during the 19th century by a succession of innovations, such as the Dutch mill, the use of iron for moving parts, automatic control (American wind wheels, etc.), and hydropower turbines. With the derestriction of flour milling in about 1850, new water mills and, above all, windmills were constructed far and wide, and numbers peaked in about 1900, when there were some 900 water mills and 2,700 windmills. This coincided with the culmination of Danish milling technology through the experimental activities of Poul la Cour. But the improved technology could not prevent wind and water power losing ground during the ensuing years to the large steam-

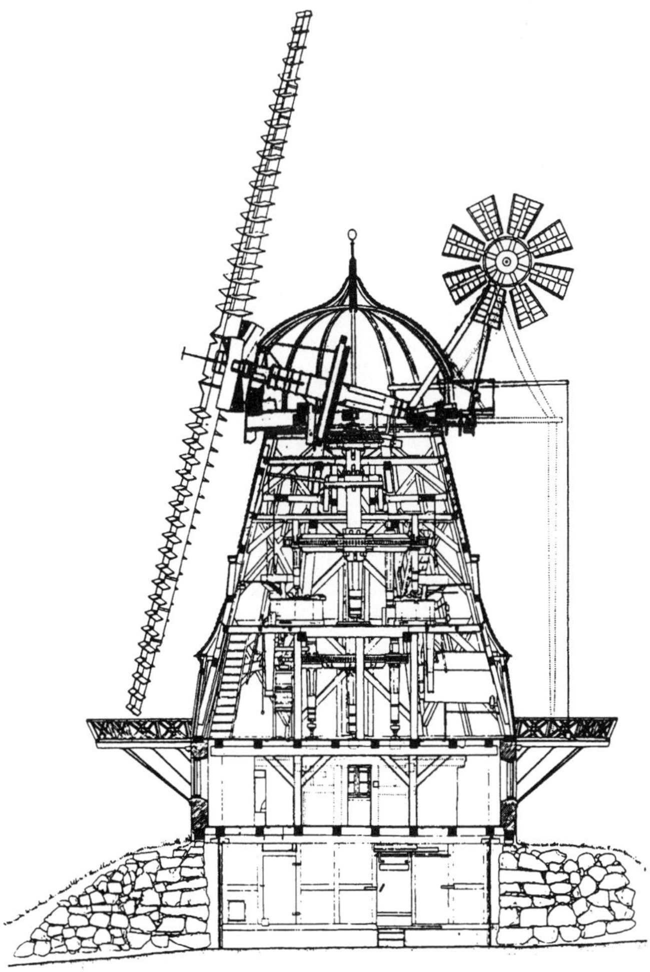

Viby windmill in Kerteminde, North Fyn. This listed Dutch windmill was constructed in 1876, at the end of the grain bonanza. Cut-away drawing of the mill with its existing machinery. The mill now belongs to the Society of Friends of Danish Mills.

Workshop building of Den Kgl. Silkefabrik ("The Royal Silk Mill"), Bredgade 36, Copenhagen. The factory was built in 1754, purportedly by the Royal builder N. Eigtved. The front building also survives and is listed. During the 19th century, these buildings housed a large steam-powered cloth mill.

Photo: Ole Hyldtoft, 1991.

powered mills and mills powered by electricity or petrol engines. The energy crisis of the 1970s, however, brought a resurgence of interest in wind power, and it is partly thanks to traditions from the turn of the century that Danish windmill constructors were able, during the 1970s and 1980s, to achieve a considerable volume of exports.

Many mills have survived, representing a wide variety both typologically and geographically speaking. This is due, not least, to the vigorous and internationally respected efforts of Anders Jespersen on behalf of the National Museum (Nationalmuseet). Before 1970, interest in industrial conservation focused exclusively on this sector. No fewer than 48 windmills and 40 water mills are listed today, not forgetting another 6 water mills and 7 windmills in open-air museums across the country.

Textile and Paper Mills

In Denmark as in other countries, the textile industry occupied a pivotal position in the early stages of industrialization. During the 19th century, scores of textile mills were established, first in the countryside where there was water power available, but then spreading, thanks to steam power, to most towns. During our own century, the textile industry has had difficulty standing up to international competition, and today it is a mere shadow of its former greatness.

Several industrial buildings from the first industrial expansion in about 1800 have been preserved in Copenhagen.The most important factories included a cloth mill, Det kgl. militære Uldmanufaktur ("The Royal Military Wool Manufacture"), established between 1760 and 1776 at Rigensgade 7-9. Its buildings, which were later turned into a military hospital, are now listed as a scheduled monument. The front building of Den kgl. Silkefabrik ("The Royal Silk Factory," 1755) at Bredgade 36, a typical, generously fenestrated manufacturing building, has also been preserved.

At the beginning of the 19th century, several cloth manufacturers shifted their production to the countryside, lured by water power and low wages. In 1810 the state transferred a substantial part of its textile production to Usserød near Hirsholm, north of Copenhagen. Today these listed buildings are occupied by a large electronics factory, S.T. Lyngs-Private. Textile manufacturers joining the exodus from Copenhagen included J.C. Modeweg, who in 1831 settled at Brede Klaedefabrik on the Mølle River, 10 km north of Copenhagen. At a nearby water mill, Stampen, which at one time was also a textile mill, the museum has reestablished an old wool spinnery from

Rebstrup in Jutland, complete with its original 19th-century machinery.

At most of the country's textile mills, water power soon became insufficient and therefore had to be supplemented with steam. At Grejs Mølle, north of Vejle in Jutland, the owner installed a 25 h.p. steam engine in 1847 and at the same time built a four-story "steam factory"—an imposing achievement at that time. This factory, now listed as one of the first generation of Danish industrial buildings, was subsequently used, among other enterprises, by a large furniture factory. A wool spinnery and steam weavery of the kind existing in virtually every town at the end of the 19th century had been set up at Den gamle By, the Århus open-air museum. Holger Petersen's factory in Tagensvej, Copenhagen, was recently listed, as an example of a large textile mill built in the 1880s in one story with shed roofs, together with homes for its workers. One of Denmark's largest steam-powered textile mills, M.K. Brandt Klædefabrik in Odense, was turned into a cultural center in the 1970s,with an art gallery, cinema, museums, and restaurants.

The many knitwear factories in and around Herning in Central Jutland are an interesting chapter in Danish textile history. This production dates back to the widespread cottage industry that flourished in the 18th century, based on intensive sheep farming and supplementing the limited agricultural output of the thin-soiled moorland holdings. The change to modern industrial production began about 1870, and today much of Denmark's textile industry is gathered in Herning and in the two small neighboring towns of Hammerum and Ikast. The local Herning Museum has a special department for the knitwear industry.

Lithograph (1888) of Holger Petersen's textile mill at Tagensvej 83-85, Copenhagen. The low, flat building with the typical shed roof dates from 1883. On the left can be seen workers' homes from 1885, laid out like a kind of "familistère."

Source: N. Malmgren, *Danmarks industrielle etablissementer*, 1, 1888.

At the beginning of the 20th century, the textile industry ran into serious problems, and several of the older factories were converted to other production. Bruunshaab near Viborg in Jutland was one of Denmark's leading textile mills during the first half of the 19th century. When cloth production here ended in 1919, the buildings were taken over by a mill-board factory. That undertaking also had to close down in 1986. The mill-board factory has now been reopened as a working factory museum. Klostermølle Pulp and Paper Mill (1873), located in Central Jutland, south of Silkeborg, is now listed together with its impressive drying shed. A great deal of the raw material here came from the local woods, and the turbines were powered by water from Gudenåen, the biggest river network in Denmark. There are still three large paper mills operating in, respectively, Silkeborg (1844), Odense (1874), and Næstved (1875), and early buildings have been preserved at all three. The Printing Museum at Brandts Klædefabrik, Odense, covers both printing itself and the closely associated paper industry. The cradle of Denmark's paper industry was on the Mølle River, north of Copenhagen, at the Strandmøllen (1643) and Ørholm-Nymølle (1793) mills. Both these mills closed down a long time ago, and the buildings, most of them from the 19th century, are now administered by the National Museum.

Flour, Butter, Bacon and Beer

Economic growth from the mid-19th century onwards was based on increasing exports of grain to Western Europe. As part of this expansion, one or more steam mills were built in nearly all Danish towns. Only one of these many plants is a listed building, namely the Skelskør steam mill, built in 1853. Many large merchant houses, complete with outbuildings and granaries, still testify to this period of prosperity. The Odder water and steam mill in Eastern Jutland is of slightly more recent date; steam machinery was installed there in 1883. Today it houses the town's museum of social history.

From the end of the 1870s onwards, agriculture was hit by heavy falls in the price of grain, combined at the end of the 1880s with German and British restrictions on imports of livestock. Faced with these serious threats, Danish agriculture in the ensuing decades underwent a thoroughgoing realignment towards two upgraded agricultural products, butter and bacon, intended predominantly for the British market. As part of this realignment, hundreds of new dairies and, a little later, scores of new export slaughterhouses were established. The continuous butter separator, invented by L.C. Nielsen in 1878 and mass-produced by Burmeister & Wain in Copenhagen, was a key technology of this industry.

The earliest dairies and slaughterhouses were privately owned, but cooperative enterprises were soon predominant. The first cooperative dairy, set up in 1882 in the country town of Hjedding in West Jutland, is now a dairy museum, fitted with the machinery and implements of the period. Another dairy, established on the island of Mand in 1897, has been transferred to the open-air museum at Hjerl Hede in Northwest Jutland, with its vintage machinery still in working order. On the other hand, although the large bacon factories are a vital part of Denmark's industrial profile, none of them has yet been listed. In Copenhagen, the public slaughterhouses from between 1878 and 1901, in the area known as Brune Kødby, are now listed, while the corresponding facility in Odense is the headquarters of TV2, Denmark's second television channel.

Denmark's famous brewing industry is better represented. Several preindustrial breweries and distilleries have been preserved, often together with large merchant homesteads in the provincial towns. Industrialization of the brewing trade gathered speed in the mid-19th century with the introduction of steam power and of lager beer. Some of these early facilities, frequently of greater architectural distinction than most factory buildings of the period, have been preserved both in Copenhagen and in the provincial towns. The two big breweries that are still operating in Copenhagen, Carlsberg and Tuborg, include many early buildings, and Carlsberg Brewery (1847) in particular has several notable industrial buildings from different phases of its development. The nearby Trekroner Bryggeri (1899) was renovated and reopened as a modern office building in 1989, following a vigorous campaign by the local preservation association. As an example of a medium-sized provincial brewery, we have the listed Bies Bryggeri (1851) in Hobro, East Jutland, while the Fjerritslev Brewery and Community Museum in North Jutland represents the small "household beer" brewery of the 19th century.

Brickworks and Cement Factories

Brick-making came to Denmark from the south during the Middle Ages, but did not become one of the main industries until the upturn of the 1840s and 1850s. The earlier technique of intermittently fired kilns is represented at Bladstrup Teglværk, dating from 1889 and reerected at the open-air museum in Odense. The new ring kiln technology with its superior fuel economics became widespread in the 1860s. The early generation of ring kilns is represented by the preserved and restored facility at Nivaagaards Teglværk north of Copenhagen, supposed to be the earliest surviving Hoffmann kiln anywhere. A more recent ring kiln forms the nucleus of a community museum at Lilleskov, southwest of Odense on the island of Fyn. An inventory of surviving brickworks was compiled in 1985–1986.

The most important concentration of brickworks in Denmark is located at Egernsund, near Sønderborg in South Jutland, where use was made of extensive deposits of stone-free glaciolacustrine clay. A large museum project is in progress here to reestablish Cathrinesminde Teglværk and its surroundings as a brick-making museum and industrial museum for South Jutland. Cathrinesminde Teglværk was founded in 1732. Brick-making in the area expanded dramatically during the 19th century, and at its peak in 1880, numbered 38 operational brickworks. This is where the world's first tunnel kiln was patented and operated in 1840, though it failed to catch on. The change to ring kilns in about 1900 and to tunnel kilns after the Second World War spelled the end of many small brickworks, but even today, eight brickworks are still active in this area.

Modern cement production belongs to the second industrial wave of the 1890s, although the first production of Portland cement started in Copenhagen in 1860. Between the end of the 1880s and the outbreak of the First World War, eight large Portland cement factories were established in North Jutland, three overlooking Mariager Fjord and five in Ålborg. Large deposits of very pure chalk and easy access to shipping transport were powerful attractions. The first of the Ålborg factories, at Rørdal, was founded in 1891 by the firm of F.L. Smidth & Co., which used it as a pilot plant, introducing, for example, M. Davidson's ball grinder and in 1899 Europe's first rotary kiln. F.L. Smidth, who started an engineering works of his own in Copenhagen in the 1890s, later established cement factories all over the world. Today Rørdal is the only Danish cement factory still in operation. None of the old shaft furnaces has been preserved, but several of the early factories have been converted for other industrial purposes. At Mariager, the old cement-making plant has been renovated to form an attractive business and enterprise estate with historic antecedents, and the former rotary kiln is once more in operation, this time for purifying polluted soil.

Old lime shafts can be visited at Daugbjerg and Mønsted, west of Viborg in Central Jutland, and open-cast mine workings in Zealand, at Stevns and Faxe. The sole surviving glassworks is Holmegård Glasværk at Næstved in South Jutland. This was founded in 1824 near a large bog that supplied the necessary fuel. The firm has established a small glass museum here, and the surviving early buildings include workers' homes from 1827–1828 and the glass house from 1874. Bornholm, a small island in the Baltic, is

The Cathrinesminde brickworks at Egernsund in eastern South Jutland. Top: oil painting c. 1870, when the brickworks operated with a double gas generator and traditional methods of production for pugging and shaping the clay. Bottom: aerial photograph, c. 1950. When the ring kiln was built in 1982, production was mechanized. In the background is the big clay pit, and in the foreground workers' accommodation, comprising four apartments.

Photos: The Museum of Sønderborg Castle.

known for its firm bedrock. Sandstone and granite have been quarried here through the centuries, and traces of this activity have left their mark on large parts of the island. Several granite quarries are still active, supplying stone, for example, for the big tunnel and bridge project between Fyn and Zealand. At Moseløkken, in the northwest of the island, a small stone museum has been set up next to a quarry that is still being worked.

Metal and Engineering

From the onset of industrialization, the metal and engineering industry was a key sector, providing machinery and equipment for other branches of industry. Two factory colonies from the end of the 18th century have survived in a fairly good state of preservation. One of them is Hellebæk Geværfabrik, a small arms factory near Helsingør (Elsinore) in North Zealand. A small musket factory was built here in about 1600 by the Crown. Passing into private ownership, it was greatly expanded during the second half of the 18th century. The listed buildings include the main building, a hammer forge, a testing shed, and many homes for workers and master-craftsmen. The other is Rådvad Isenkramfabrik on the Mølle River, north of Copenhagen. Between 1759 and 1972, millions of hardware articles were produced at this water mill, most of the time under the aegis of Københavns Isenkræmmerlav. Some buildings, such as the distinctive file factory, date from the mid-18th century, but most were built during the expansion of the 19th and early 20th centuries. Today the listed industrial area houses the Nordic Center for the Preservation of (Early) Crafts, i.e., crafts that are on the point of dying out and are needed for the future restoration of buildings.

Burmeister & Wain's large diesel engine assembly plant at Christianshavn, Copenhagen, built in 1923–1924. Demolished 1991.

Photo: Ole Hyldtoft.

Up till about 1840, Statens Værksteder at Frederiksværk was the leading iron-manufacturing and engineering enterprise in Denmark. Frederiksværk in North Zealand is one of the few Danish downs to have sprung up round a single main industrial complex. The town's history goes back to about 1750, when a bronze foundry was established, driven by water power from a canal dug by Swedish prisoners of war from Arresø to Roskilde Fjord. At the end of the 18th century, the proprietor expanded the operations to include a powder mill, a copper rolling mill, an iron foundry, and an engineering works. Of the old operations, the powder mill is in a good state of preservation. The old, timber-winged foundry building from about 1770 was recently reinstated and much of it rebuilt as a cultural center.

During the 1840s and 1850s, many iron foundries and engineering works were established in the capital and in most of the provincial towns. The listed Køge Jernstberi (1842–1875), south of Copenhagen, is one example of a small provincial foundry of this kind, and provincial towns all over Denmark retain buildings from this first generation of iron foundries. At the end of the 19th century, some of these foundries established nationwide production and distribution of tiled stoves and kitchen ranges. They included, for example, N.A. Christensen in Nykøbing Mors, North Jutland, which has set up a small foundry museum. A large collection of tiled stoves can be seen at Den gamle By, the open-air museum in Århus.

From the mid-19th century onwards, the technical lead in Danish engineering was taken by Baumgarten & Burmeister, later Burmeister & Wain. This firm's main output consisted of industrial machinery, steam engines, and steamships. B & W acquired an international reputation at the beginning of the 20th century when in 1904 it began producing stationary diesel engines, followed in 1912 by its first ocean-going motor vessel, the *Selandia.* A good deal of international motor vessel tonnage then came to be powered by diesel engines built by B & W or their licensees. The company has a museum of its own in Copenhagen.

The two big industrial museums, Danmarks Tekniske Museum in Helsingør, north of Copenhagen, and the Industrimuseet in Horsens, East Jutland, both have large collections of engines, working machinery and other products of the Danish engineering industry.

Infrastructure

The establishment of efficient public supplies of water, gas, and electricity played an essential part in the industrialization process. The numerous public facilities have only attracted limited attention. With just one exception, none of these often highly interesting buildings has been listed, but many waterworks and power stations have survived in new capacities. The surviving waterworks include the buildings of Copenhagen's first steam-powered waterworks 1857–1859, now converted into a cultural center. Gas played an important part in Denmark for the hundred years between 1850 and 1950, but despite several attempts it has not been possible to establish a gas museum and preserve one of the old coal-fired gasworks. The only listed gasworks monument is Martin Nyrop's monumental brick gas holder at Østre Gasværk (1881–1883) in Copenhagen, which has now been turned into a theater.

Electricity arrived with the industrial upturn of the 1890s. Teknisk Museum in Helsingør has an important collection from the infancy of electricity, which in Denmark goes back to Hans Christian Ørsted's discovery of electromagnetism in 1820. In 1984 the Tange (Bjerringbro) Electricity Museum was opened in Central Jutland, next to Denmark's biggest hydroelectric power station. This plant, still operating with three Francis turbines of 2,000 h.p. each, was commissioned in 1921. A new exhibition building also includes a demonstration of electricity production with a reciprocating steam engine, a steam turbine, and a diesel engine. The Horsens Industrial Museum occupies the town's old power station, built in 1907, and at Randers in East Jutland, the big power station from 1906 has just been reopened as a concert hall and theater.

Denmark's shipping traditions are perpetuated by a number of museums. The Danish Maritime Museum is housed at Kronborg Castle in Helsingør (Elsinore). Local shipping and fishery museums exist, for example, at Marstal on the island of Ærø, south of Fyn, and in Esbjerg, Hirthals, and Grena, all of them in Jutland. Denmark's navy, once one of the biggest in Europe, recently acquired a museum of its own, Orlogsmuseet in Copenhagen. In addition to exhibits and models, the museum also has a unique collection of blueprints. The paddle steamer *Hjejlen,*

The Horsens Museum of Labor, Crafts, and Industry occupies the town's old power station, built in 1906 by architect Viggo Norn. It is in Art Nouveau style, with checkered flagstone flooring, granite door surrounds, decorations under the ceiling, and about 1,600 window panes. The industrial museum opened in 1977 is intended to illustrate the transition from the agrarian, market-town community to industrial society.

Photo: Hans Vollertsen, The Industrial Museum at Hortsen.

built by Baumgarten & Burmeister in 1862, still serves the Silkeborg Islands in Central Jutland. Other ships preserved include the frigate *Jylland* at Ebeltoft, East Jutland, and the diesel-powered rail ferry *Sjælland* (1951), now converted into a museum and restaurant and permanently moored in the Port of Copenhagen.

A Danish road museum opened in 1991 on the island of Farø, with the opening of the new motorway bridge between Zealand and Falster. There are three transport museums in Billund, West Jutland, otherwise known for its LEGO factory. They are Danmarks Flyvemuseum (aviation), Falck Museet (rescue and ambulance), and a motor museum. There is a fine collection of early Danish cars on display at Teknisk Museum, Helsingør. And there are private motor museums in Ålholm, near Nysted on the island of Lolland, in Egeskov, on the island of Fyn, and in Gjern, in Central Jutland.

The Hammelcar was the first Danish automobile, probably from around 1890. The photo shows the constructor Hans Urban Johansen in the car in the courtyard of Hammel's factory, Nørregade 38-40, Copenhagen. The car, still in working order, is now on display in the National Museum of Science and Technology in Helsingør (Elsinore). *Source:* The National Technical Museum in Denmark.

Railway Stations

The first railway in Denmark proper, from Copenhagen to Roskilde, was opened in 1847. Between 1857 and 1875, a nationwide rail network was established, and the next and largest expansion of Denmark's railways came between the mid-1890s and the First World War, with the establishment of a host of inland lines. Many years' teamwork by the Danish Research Council for the Humanities, the National Planning Agency, and DSB (Danish State Railways) has resulted in a detailed inventory of DSB buildings and facilities all over the country. In the 1980s, DSB followed up this work by preserving and restoring railway stations on Kystbanen (the coastal line) from Copenhagen to Helsingør, and in 1991 these efforts culminated in many new listings, so that preservation orders now exist for 39 railway facilities.

These listings include examples of various styles and periods, from the Germanic Roskilde Station (1847) via the more anonymous English style from about 1860, the heyday of the Romanesque style in the 1860s and 1870s, the Renaissance revival of the 1890s, and the Cottage style of about 1900 to the Functionalist Nørrebro Station built in 1930. Both more notable town stations and a number of smaller rural stations are represented, and often the listings include, not only the main building but also several outbuildings and installations, such as engine sheds, water towers, signal boxes, warehouses, and toilets. This is the case, for example, with Guldager, a country station in West Jutland dating from 1874.

By contrast, the many stations on small private lines have been virtually overlooked. Here there are no systematic inventories, and only a few stations, e.g., Åge Rafn and Kays Fiskers' four unique rural stations (1915–1916) on the Gudhjem-Almindingen line on the island of Bornholm, have been preserved for posterity. DSB has its own railway museum in Odense. Several private associations intermittently operate their own "historic" lines, e.g., in Helsingør, Bandholm (Lolland), and Bryrup (Central Jutland).

The Copenhagen Waterfront

The biggest current preservation problem concerns the Copenhagen waterfront. Just as in other European seaports, modernization and the introduction of container traffic have deprived large areas and many old buildings in the Port of Copenhagen of their original functions. In addition, many of the big, old industrial concerns in the port, such as the Burmeister & Wain diesel engine factory and De Danske Sukkerfabrikker (sugar), have moved their production facilities away from the dockside. These problems have been compounded by the decision to close the neighborhood dockyard and to transfer its activities to Korsør and Frederikshavn. The many valuable buildings and facilities that have been endangered as a result include highly interesting warehouses from the 1890s in Frihavnen (the free port); many unique buildings at Holmen, several of which are listed and date back to about 1750; Burmeister & Wain's large diesel engine plant from 1923–1924; a counterpart of Peter Behren's famous workshop in Berlin, H.C. Ørsted Værket, with an 8-cylinder 22,500 h.p. B & W diesel engine from 1932; and the Ford assembly plant in Sydhavnen, designed by Albert Kahn and constructed in 1923–1924. In 1988 the National Planning Agency published an extensive inventory of buildings and facilities in the Port of Copenhagen.

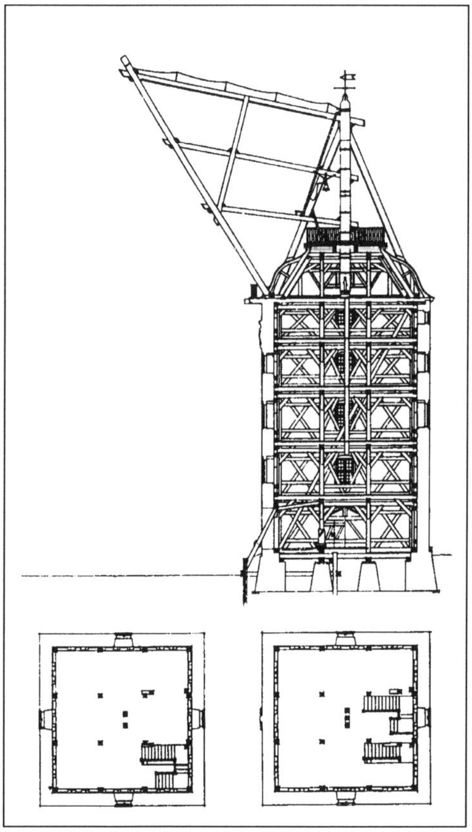

One of the most conspicuous landmarks in the Port of Copenhagen is the large rigging shears at Nyholm, built in 1748–1751, by the architect Philip de Lange. This listed and well-preserved installation consists of two separate structures, a solidly anchored crane structure of timber and a rebuilt granite rubble edifice with a copper roof. *Source:* Chr. Elling and V.S. Møller, in *Holmens bygningshistorie 1680–1930*, 1991.

Already in the 1970s, many old warehouses in the inner part of the port, which date from the commercial prosperity of the end of the 18th century, were assured of preservation by being turned hotels, apartments, and conference and exhibition centers. Several warehouses in Frihavnen have been lost, but a local preservation plan is aimed at preserving a couple of the most outstanding of those that remain. At Holmen, preservation orders have been requested for another 15 buildings, including the distinctive, monumental facility constructed when iron ships began to be built in the second half of the 19th century. If preservation orders are obtained, this will secure the main elements of the overall facilities. In addition, there are plans for a large museum project that will include a repair yard for veteran ships, a harbor museum, and a holiday center in part of the old facilities. Despite a major preservation campaign, much of Burmeister & Wain's production facilities were demolished in 1991, including the diesel engine assembly plant from 1923–1924. In Sydhavnen (South Docks), preservation orders have been requested for the large diesel engine at H.C. Ørsted Værket and also for the Ford factory.

Housing

Any number of workers' homes are still being used as such. The extensive renewals that began in the 1970s often involved brutal treatment of old housing developments, but the tendency during the 1980s favored sympathetic renewal. A number of workers' homes in the old rural industrial communities are now preserved and listed. Statens Nyboder, in the center of Copenhagen, originally constructed for the permanent employees of the navy and naval dockyard, is both unique and listed. This distinctive colony of low-rise enfilades containing small apartments was constructed in the 17th century and subsequently enlarged.

Otherwise not much industrial housing is listed, and the exceptions are not typical. They include the first philanthropic working class housing development in Copenhagen, two blocks erected in 1852 (at Overgaden oven Vandet 66-74) by a specially constituted company. A much larger development is Lægeforeningens Boliger, built between 1853 and 1872 on what was then the outskirts of the city. Both developments were designed by eminent architects. Also listed are the fine terrace-house development of Bakkehusene at Bellahøj, by Th. Henningsen and Ivar Bentsen, built by a housing cooperative in 1921–1922, and Kay Fisker's impressive Hornbækhus courtyard development, built by a housing cooperative in 1923; both are located in Copenhagen. Workers' homes are one of the main concerns of Arbejdermuseet in Copenhagen. Nationalmuseet has also concerned itself with this field, and apartments from a workers' housing cooperative of the 1890s form part of the Horsens Industrial Museum.

Preservation

The state has been empowered since 1918 to list selected secular buildings for preservation. The prospects of preserving industrial buildings were strengthened in 1979 with the passing of the new Preservation of Buildings Act, to protect early buildings of architectural, historic, and environmental interest, including buildings that illustrate housing, working, and manufacturing conditions. This law is administered by the National Agency for Physical Planning, conjointly with owners, local authorities, and county museums. The new opportunities are yet to be fully utilized. Only about 10 percent of the 3,600 buildings listed are connected with industrialization, and most of these are warehouses, water mills, and windmills.

In addition, since 1960 many country towns, acting in collaboration with the National Museum and the National Agency for Physical Planning, have compiled inventories and preservation plans. These efforts have been followed in recent years by the publication of many inventories in the form of "town atlases," aimed at drawing the attention of owners to essential architectural values and also for serving as a basis for local preservation plans. During the 1970s and the early 1980s, with support from the Danish Research Council for the Humanities, a nationwide inventory was compiled of Danish factories (National Museum). Several national and some local associations are involved in the preservation of industrial buildings and facilities. This work is being partly coordinated by the Danish Association for the Study and Preservation of the Industrial Heritage, which among other things applies for preservation orders, arranges meetings, and, for a number of years now, has published the journal *Fabrik og Bolig*.

The work of preservation accelerated during the 1970s and 1980s. Apart from work relating to windmills and water mills and the two long-established, nationwide museums, Nationalmuseet and Teknisk Museum in Helsingør, practically all the above mentioned preservation and museum initiatives have come during the past 20 years, and most local museums of social history today are mindful of Denmark's industrial heritage. Previous efforts have been further promoted in recent years by an incipient change of attitude and a heavy decline in building development. Two of the new museums give wide coverage to industrial buildings and the people who worked and lived in them. One of these is the Industrial Museum in Horsens, opened in 1977, and the other is Arbejdermuseet in Copenhagen, which in 1983 opened its first exhibition of working life in the 1950s, in buildings that for generations have been the bastion of the Danish labor movement.

There have been setbacks, however. In the mid-1970s, despite a major campaign, it proved impossible to save one of Copenhagen's most distinguished industrial buildings, the Marstrands Mølle steam brewery from 1865 (later known as Kongens Bryghus). In the mid-1980s, the same fate overtook one of Denmark's first factories to incorporate ferroconcrete on a considerable scale, the monumental building of De Danske Cikoriefabrikker in Valby, Copenhagen. Almost simultaneously, one of the country's oldest textile mills, the renowned Randbøldal Klædefabrik near Vejle in East Jutland, disappeared as a result of the proprietor's neglect of the buildings. In 1990, the Ny Geværhus small arms factory in Copenhagen was demolished as a result of top-level political pressure. That factory dated from about 1890 and was notable both for the typical Romanesque style of the period and as the first Danish factory to make consistent use of American methods of production. The site remains undeveloped. The same goes for the next major victim, the above-mentioned Burmeister & Wain diesel engine plant from 1923–1924, demolished in 1991, which was not only of outstanding architectural interest but also a symbol of early Danish technology of international standards.

Further Reading

Jørgen Sestoft: *Arbejdets bygninger* (1917).

Poul Strømstad(ed.), *Mennesket og maskinen* (1980).

Axel Nielsen (ed.), *Industriens historie i Danmark*, I–III (1943–1944).

Richard Willerslev, *Studier i dansk industrihistorie 1850–1880* (1952) (Ph.D. thesis).

Sv. Aa. Hansen, *Early Industrialisation in Denmark* (1970).

Ole Hyldtoft, et al., *Det industrielle Danmark 1840–1914* (1981).

Ole Hyldtoft, *Københavns Industrialisering 1840–1914* (1984).

Hans Chr. Johansen, et al., *Fabrik og Bolig i Odense 1840–1940* (1983).

Henning Bender *Aalborgsindustrielle udvikling fra 1735 til 1940* (1987).

Jørgen Burchardt, *Provinsindustri: Arbejdspladser på Faaborgegnen de sidste 100 år* (1984).

H.R. Lauridsen, *Viborgs industrihistorie 1742–1990* (1990).

Jeppe Tønsberg, *Lyngby-Bogrn* (1984).

See also the journal *Fabrik og Bolig*.

Brede Klaedefabrik

Ole Hyldtoft

Brede Klaedefabrik textile mill is on the Mølle River, about 15 km north of Copenhagen. The Mølle river was often referred to as the cradle of Danish industry. This little river, 12 km long from its source in Furesøn to its estuary in Öresund (the sound between Denmark and Sweden), and with a total fall of only 20 meters, drove nine watermills that, for nearly 300 years played a pivotal part in Denmark's industrialization. For example, Scandinavia's first paper-making machine was installed at the mill furthest downstream, Strandmøllen, in 1829. This intensive exploitation of the river was prompted first and foremost by its proximity to the Copenhagen market and to the financial and commercial resources of the national capital. The mills began as grainmills in the Middle Ages, but by the 16th century, most of them had been converted to industrial use, producing gunpowder, copper, muskets, textiles, and paper. Today nearly all industrial operations have been discontinued, and four of the mills have been taken over by the National Museum. Today the Mølle River, with its hilly, wooded surroundings is one of the most important recreation areas for the capital, with numerous surviving relics of the early industrial history of the region.

Brede klædefabrik–textile mill. The old factory building from the 1850s from the pond.

Brede Klaedefabrik is the largest industrial facility on the Mølle River. In the 17th century, the local grainmill was converted, first into a powder mill and later into a copper and brass factory. In 1831 J.C. Modeweg transferred his cloth mill from Copenhagen to the water-powered plant at Brede, where in the following decades, he built up one of the country's leading textile factories. Already in 1842, the local water power had to be supplemented by a steam engine, and in the 1850s, the old waterwheels were replaced with turbines. Subsequent expansions culminated in about 1900 with the establishment of a grandly conceived, patriarchal welfare community with its own dining hall, general store, nurs-

The spinnery, built in ferroconcrete, from 1908.

ery, school, asylum, water works, gasworks, and electricity plant, and numerous living quarters for workers and foremen at the factory. In 1890 one of the loom sheds was equipped with Denmark's first automatic sprinkler installation, and in 1908, the factory opened a new spinning shed, the first allferro-concrete industrial building in Denmark.

The mill closed in 1956, and shortly afterwards, it was taken over by the National Museum, the third (Modern) department of which now uses the factory buildings, etc. for exhibitions, workshops, storage facilities, and offices. The old factory environment is quite intact. The oldest part is the main building from 1795, while the other buildings are best described as a catalogue of typical phases in Danish factory and housing construction from 1840 to 1920. Only the main building has been listed. Owing to the financial straits of both government and the cultural sector today, parts of this complex may be endangered, and efforts are therefore being made to secure protection for the complete environment, since its completeness is what makes it unique.

The Old Carlsberg Brewery

Ole Hyldtoft

Brewing was one of the earliest industries to be mechanized and one of the most capital-intensive. At the same time, breweries are consistently more architecturally expressive than most other factories. The most successful of the new lager breweries was Carlsberg. Its founder, J.C. Jacobsen, inherited an old small-beer brewery at Brolaeggerstraede 5 in the center of Copenhagen, the buildings of which are still extant. When, in the mid-1840s, Jacobsen decided to go in for the new "Bavarian" beer, he chose to erect a completely new factory at Valby, a couple of kilometers west of the city and close to the new railway line.

The new brewery was opened in 1847 under the name of Carlsberg. The beer caught on, and through the period ending in 1879, Old Carlsberg was expanded into a large industrial complex with extensive production, warehousing, and stabling facilities, a main building, complete with parkland, forming the owner's private residence, and numerous living quarters for workers and foremen at the brewery. The buildings were of restrained, sober classical design, the work of two of the most eminent architects of the period, H.C. Stilling and N.S. Nebelong. At the same time, the intention was for the brewery to keep

The New Carlsberg Brewery, with the famous Elephant Tower and the Brewery, both buildings from 1901.

The Old Carlsberg Brewery from the 1840s.

abreast of the latest technical requirements. It was steam-powered from the beginning, and its rebuilding after a fire in 1867 marked an innovation in Danish industrial architecture: the use of rolled steel sections for floor structures.

Regular production at Old Carlsberg came to an end in the 1920s. For sentimental reasons, the company decided against dismantling and demolishing the facility, and instead it was put in mothballs. Consequently, when the company decided at the end of the 1970s to reestablish Old Carlsberg as a brewing museum, it had an excellent foundation to build on. The plant includes several generations of machinery and brewing equipment from the 19th century, and when it was reopened, efforts were made to revive the atmosphere of the old brewery by putting machinery, motors, and other apparatus back into their original working order. The former main building is now a grace and favor residence for an eminent Danish scientist.

At the same time, Old Carlsberg is an early example of a closer link between science and industry. J.C. Jacobsen himself attended lectures at the Copenhagen Polytechnic in the 1830s, and he was always alive to the implications of scientific progress for the brewing industry. As early as 1871, he fitted out a small brewing laboratory at Old Carlsberg (this now forms part of the museum), and in 1875 he founded, for primary research, the Carlsberg Laboratory, today known as the Carlsberg Research Center, which under the direction of Johan Kjeldahl and Emil Chr. Hansen rapidly acquired an international reputation through Kjeldahl's method for determining nitrogen in organic substances, and for Hansen's isolation of yeast strains.

Both Old Carlsberg and the succeeding facility at Valby are remarkable industrial monuments. Unfortunately, the brewery annex, erected between 1871 and 1876, was recently pulled down. On the other hand, large parts have survived of the impressive and somewhat eccentric New Carlsberg, built 1880–1903 for Carl Jacobsen, the son of J.C.

The Carlsberg Brewery. The steam engine from 1879 is still in its original place.

Jacobsen, and designed by the architects Vilh. Dahlerup, Hack Kampmann, and Vilh. Klein. The surviving structures here include the Dipylon, the Elephant Tower, and the twisted chimney. Large new additions in neoclassical style followed in the 1920s, with Carl Harild as architect, and the latest additions in international style include, for example, the bottling plant (1967–1969), designed by Sv. Eske Kristensen.

Nivaagaard Brickworks

Ole Hyldtoft

Brick-making was one of Denmark's biggest industries. Using local moraine clay, the Nivaa district, near Öresund, about 30 km north of Copenhagen, developed from the 1850s onwards to become one of the principal brick-making enclaves, with no fewer than three important brickworks: Nivaagaard, Niverød and Sølyst.

The history of the oldest of these brickworks, Nivaagaard, can be traced back to a peasant brickworks that from 1701 onwards supplied large quantities of building materials to the royal residences in North Zealand. Later, though, this brickworks was idle for long periods of time, and it was only the rise in demand from the 1840s onwards that laid the foundations of more industrialized output. In 1854 the brickworks attained an annual output of about one million bricks, and three years later, in 1857, Nivaagaard became one of the first Danish brickworks to install a steam engine for pugging clay.

Nivaagaard Brickwork. The ring kiln from the 1870s.

The worst bottleneck in traditional brick-making, though, was the immense quantities of fuel consumed by the old, periodic chamber kilns; most of the energy supplied was quickly lost in the cooling process that followed every firing. The tunnel kiln developed in about 1840 by a Flensborg brickmaker, Hans Jordt, and M.H. Holler was an early attempt to solve this problem, but at this stage the tunnel kiln could not be made a paying proposition. Instead the breakthrough of a new, fuel-saving kiln came with the ring kiln patented in 1858 by the Germans F. Hoffmann and A. Licht. The ring kiln reduced fuel consumption by two-thirds and at the same time made firing a continuous process. In addition, the heat from the kiln could be used for drying the unfired bricks.

The first Danish ring kiln was erected, in the winter of 1866–1867, at G. Kähler's brickworks in Korsør on Zealand. Three years later, in 1870, Nivaagaard also went over to ring kiln operation. Local tradition has it that the Nivå ring kiln was built to original drawings by F. Hoffmann. It makes an imposing edifice. The kiln is 38 meters in diameter and has 15 chambers, each holding 21,000 bricks. A 38-meter

Nivaagaard Brickwork. The setting of unburned bricks inside the kiln.

chimney rises from the center of the furnace. The investment was a success, and output rose to several million bricks a year. The kiln remained almost continuously in operation for 100 years, from 1870 until 1967, so that all in all it fired practically one billion bricks.

Today the kiln at Nivaagaard Brickworks is the oldest surviving ring kiln in Denmark and probably in the world. It is listed as a historic monument, and a local institution has carried out restoration work to save it from disintegration. The aim, using the ring kiln as nucleus, is to preserve a complete brickworks—living quarters, drying plant, machinery, conveyer equipment, and tools—to illustrate the contribution that brickmaking made to society for more than 100 years. At the same time, the brick industry is a field in which Danish technology has held its own internationally, partly through the F.L. Smidth & Co. engineering company, which has established brick and cement works all over the world.

Laegeforeningens Boliger

Ole Hyldtoft

Starting in the mid-19th century, the industrialization process led to heavier concentrations of population and, consequently, to more conspicuous housing problems. By far the greater part of the new industrial working population had to make do with the primitive accommodation offered by the private housing markets, but middle-class philanthropic circles were soon trying to establish alternatives. Laegeforeningen's living quarters in Copenhagen were the first philanthropic attempt on a grand scale to remedy the housing shortage in Copenhagen at that time and to provide cheap, healthy homes for the working and unpriviliged classes.

The immediate cue for this venture was the

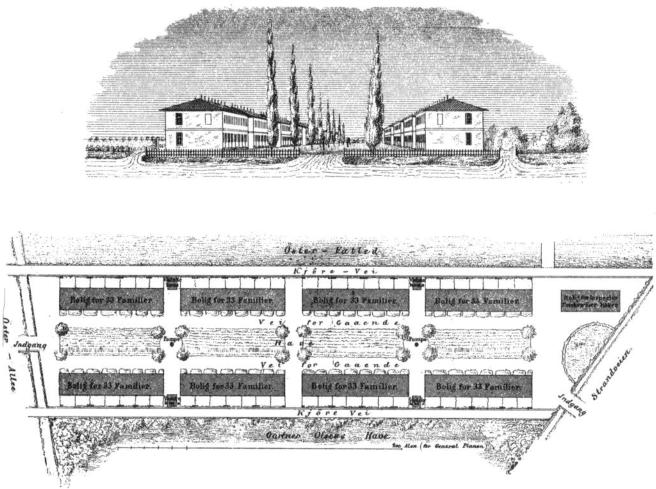

Lægeforeningens Boliger. The prospect from 1853 with view, situation plan, and the organization of the apartments.

cholera epidemic of 1853, which in a few summer months claimed more than 4,700 lives in Copenhagen. The initiative for the new housing came from Laegeforeningen, an association of physicians set up in 1853 to fight cholera. Most of the necessary capital came from private sources and was raised by an appeal for donations while the epidemic was raging. A site was found at Østerbro, on the outskirts of Copenhagen, and it was here, between 1853 and 1857, that the association built eight tenement blocks, each comprising

Lægeforeningens Boliger. Wash-houses and communal drying places.

31 small flats and an inspector's quarters, as well as shops and living quarters for the shopkeepers. This was an impressively large building scheme by Danish standards. Already in 1857, the colony had more than 900 residents, and following the addition of another ten blocks between 1867 and 1872, its population in the 1880s exceeded 2,500.

The first development was designed by M.G. Bindesbøll, internationally known for the Thorvaldsen Museum in Copenhagen, erected a year or two earlier. He designed the colony as an open, low-rise settlement of eight long, two-storey apartment blocks. Each block was surrounded by gardens for the inhabitants, and generous open and communal areas were provided between the blocks. The big expansion in about 1870, designed by Vilh. Klein, adhered on the whole to the original concept, though with slightly larger dwelling units. Later on, various institutions were added to the colony, in each of its four corners, viz., a kindergarten in 1886, a public baths in 1891, a parish hall in 1897, and handicraft facilities in 1906. Laegeforeningens Boliger did a great deal to influence subsequent developments of a similar character, and Bindesbøll's plain, unpretentious architecture continued to inspire Danish architects well into the present century.

Today Laegeforeningens Boliger forms an enclave in the middle of Copenhagen, with its own streets and with an architecture and greenery quite different from those of its surroundings. Already in the 1930s, plans existed to pull down the flats and replace them with new, more intensive development. The colony was listed in 1959, but this has not deterred the present owner, the Municipality of Copenhagen, from persistently putting forward plans for demolition. These plans have intermittently resulted in violent clashes between the authorities and residents and sympathizers. The main results to date seem to have been a gradual and sympathetic restoration of the whole settlement, which is a well-preserved and unique monument to efforts in the period concerned to create a non-profit alternative to the private housing market.

Arbejdernes Forsamlingsbygning

Ole Hyldtoft

There was more to industrialization than just factories and housing. The workers soon began organizing themselves industrially and politically, and it was not long before the powerful Danish trade union movement became a prime factor in social development. From modest beginnings in the 1840s and the 1850s, the labor movement achieved an almost explosive breakthrough during the boom period of the early 1870s. Despite fierce resistance from employers and authorities, nearly a quarter of Copenhagen's male workers had joined the new organizations by the middle of that decade. The movement expanded during the decades that followed, and in about 1900, something like 85 percent of male workers and 30 percent of female workers were members of the now officially recognized unions. It was also at this time that social democracy achieved its political breakthrough.

The first building of the trade union movement, Arbejdernes Forsamlingsbygning at Rømersgade 22, Copenhagen, was officially opened in 1879. Owing to official persecution, the workers had difficulty finding places to hold their meetings, so they decided to take matters into their own hands. The funds for the plot and the necessary building materials were raised by issuing thousands of low-denomination shares, and the actual building

The Workers Meeting House, built in the 1870s. The building is today the Labor Museum.

work was done on a mutual basis, mostly on Sundays. The position, on the site of the old ramparts, was a central one. The street front of the building was decorated in the best middle-class style of the time with an abundance of Renaissance motifs in stucco. Inside were variously sized meeting rooms, trade union offices, apartments, and, in the basement, a restaurant. The heart of the building was the large banqueting hall at the back, with room for up to 1,000 people. For the 100 years that followed, the building was an important center of the labor movement in Copenhagen and Denmark, as a setting for meetings, general assemblies, trade union work, social occasions, and festivals. During the International Socialist Congress in Copenhagen in 1910, Rosa Luxemburg made a speech to Denmark's Social Democrats in the banqueting hall. Her distinguished audience included Lenin, Jean Jaurés, Alexandra Kollontay, and Karl Liebknecht.

In 1982 the building was donated to the newly established Labor Museum, and the following year, it was listed as a historic building. The Labor Museum exists to show how Danish workers have lived and worked during the past 100 years, their housing conditions, working conditions, leisure pursuits, and their industrial and political campaigning. The museum's first exhibition opened on May 1, 1983. In addition to various minor exhibitions, the museum now has three permanent ones. The first of these illustrates the everyday life of early industrial workers (1870–1890), the second housing and working conditions during the lean years of the 1930s, and the third the Danish working-class family of the 1950s. In addition, with support from the National Museum, the Labor Museum has undertaken a careful and sympathetic restoration of the impressive, historic banqueting hall, restoring it to its appearance in 1916. Finally, the basement below the front building reopened recently as a restaurant, in the guise of a café and beer hall from 1892, while a room to one side has been fitted out as a mass kitchen from the 1930s.

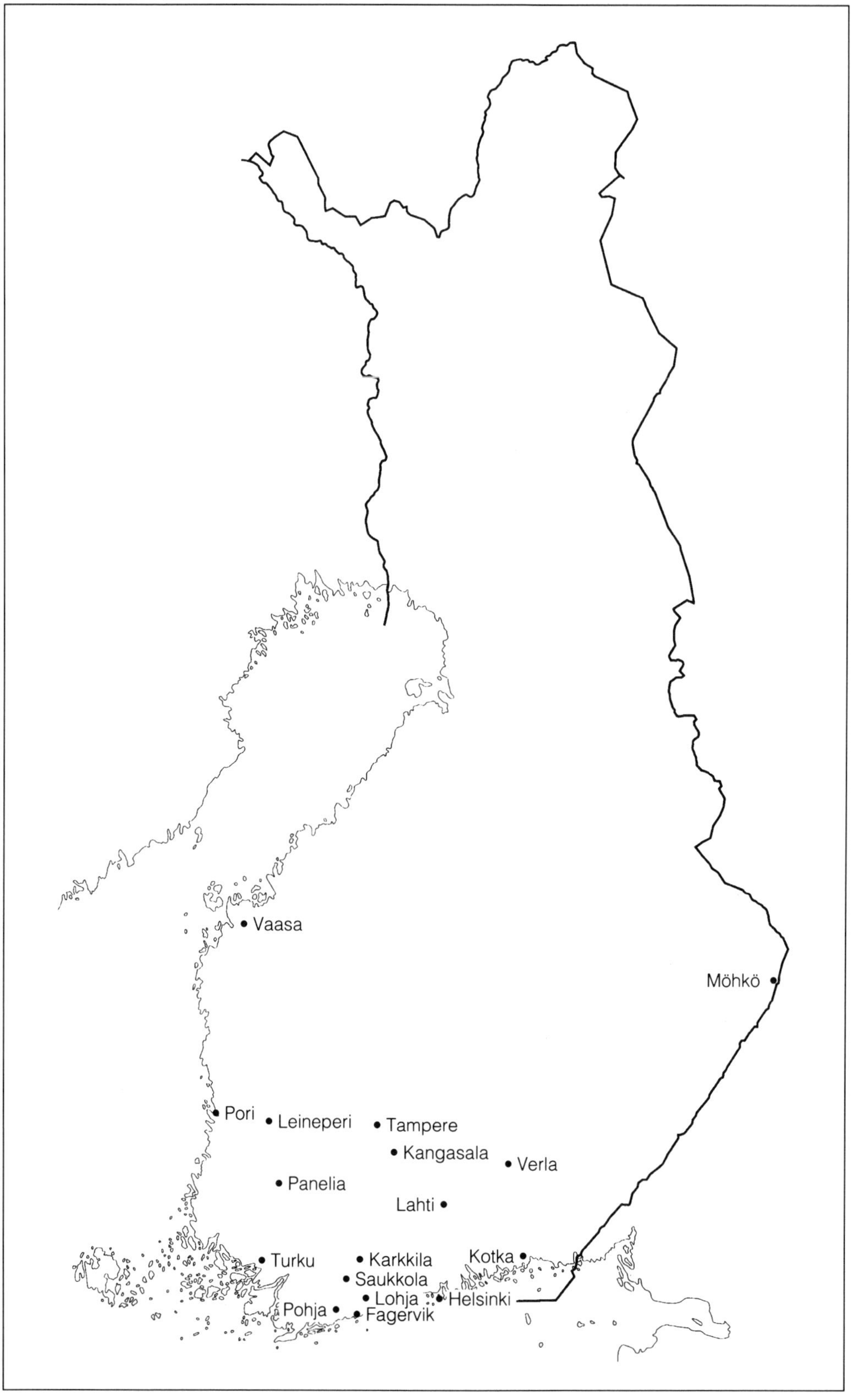
Vaasa
Möhkö
Pori
Leineperi
Tampere
Kangasala
Verla
Panelia
Lahti
Turku
Karkkila
Kotka
Saukkola
Lohja
Helsinki
Pohja
Fagervik

CONSERVATION OF THE INDUSTRIAL HERITAGE

Finland

Elias Härö **Lauri Putkonen**

A Background and an Overview

October 1991 saw the completion in Tampere, the Finnish city with the most significant industrial history, of an inventory designed to find out the number of smokestacks from the city's industrial history that are still standing within the city area, and how many are still in use, and how many should perhaps be conserved as monuments of industrial history. The inventory was accompanied by commentaries by experts presenting their views on how difficult and expensive it would be to preserve as monuments unused, cold factory smokestacks made of brick. The case is characteristic for the history of industrial milieus even in a broader sense. Tampere's industrial past spans some 160 years and coincides with a period during which Finland moved from being an agrarian to an industrial society and subsequently an increasingly urbanized, modern service society. The disappearance of smokestacks' historical industrial milieus or their conversion into items of museum interest is part of a broader change in production technology and capital structure, as well as marking a compliance with society's new norms on pollution. At the same time, the phenomenon is a serious challenge to those responsible for research into and documentation of industrial heritage

For centuries, Finland was economically an agrarian country on the European periphery, although since World War II, its industrialization has been extremely rapid. The predominantly agricultural past and the heritage of a class society attendant on it dictated almost exclusively the direction of research into cultural history, museum activity, and the conservation of buildings, so much so that industrial culture in both its physical and spiritual manifestations was seen more as a threat to those aspects of culture that had been traditionally considered valuable. Gradually, as a result of a change in attitudes in the Western industrial countries, interest was also awakened in Finland in the values inherent in industrial culture. In this context, the British industrial archaeology movement provided useful examples even though as such it did not gain a particularly strong foothold in Finland or in the other Nordic countries.

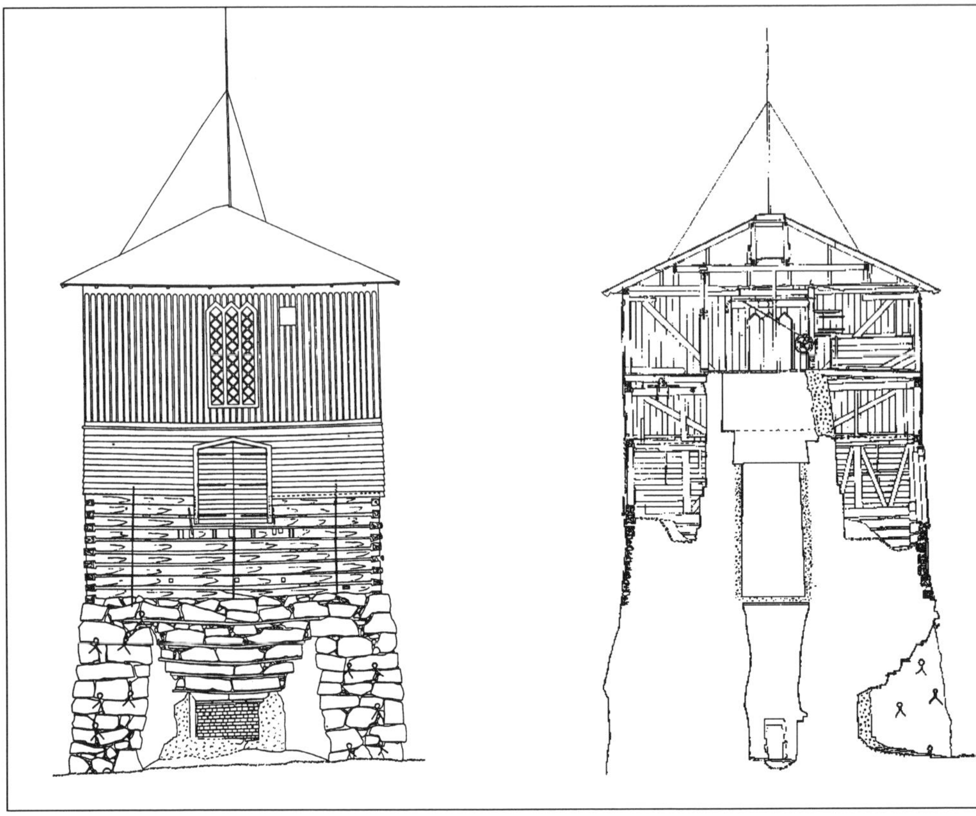

The ironworks of Leineperi, Kullaa: the northern facade and section of the timber-clad construction of the blast furnace. *Source:* National Board of Antiquities and Historical Monuments, 1989.

Against this background, the importance of raising public consciousness and providing sufficient information about the cultural values in the industrial heritage is obvious. As regards building protection, the voluntary conservation undertaken by property owners will continue to occupy a central role, and it is in this direction that the main stress of the work should lie, both of the authorities and the conservation of monuments—through research, professional organizations, amateur societies of regional history, etc. Particularly in the context of industrial culture, the problem seems even today to be the scant appreciation shown for the significance of the industrial milieus, as compared to the appreciation for the traditional, agriculturally dominated countryside or for the older urban culture. In this respect, both researchers and those who shape the public opinion still have a lot to do.

Although Finland is today primarily known as a country with a varied wood-processing industry, its oldest pre-industrial monuments represent the iron industry. As a result of expansion in the Swedish mining industry, 15 ironworks using Swedish ore were founded in Finland by the end of the 17th century. During the 19th century, when Finland was a Russian grand duchy, more than 50 ironworks were founded to process ore excavated from Finnish lakes or bogs. Understandably, these works were the first to capture public interest and to become objects of research, documentation, and protection. The older, Swedish-founded ironworks' milieus became objects of a particularly wide

The ironworks of Leineperi, Kullaa: the damaged timber construction of the blast furnace was renewed, and the surrounding casting house was reconstructed. On the left, the gable of the forge, whose foundation was reinforced to avoid collapse.
Photo: Pentti Pere, Satakunnan Museum.

interest in that they also contained manor houses and parks with cultural values. However, it was only in 1984 that government grants submitted the restoration and conservation of the two tilt hammer forges at Fagervik dating back to the latter half of the 18th century. Thus, at the last possible moment, it became possible to save a document and valuable memento of the iron industry that was threatened with almost total destruction. Towards the end of the 1980s, certain other monuments of the historical iron industry have been conserved, including one that in its time was among the greatest producers of lake ore, the Möhkö Works, situated in Finnish Karelia.

A particular difficulty concerning factories from the industrial era is that traditional methods used to protect cultural monuments are rarely practical. In Finland, as in other modern industrialized countries, the updating of technology and the rationalization of production have led to the relocation of factories, and their original premises, often extremely valuable sites, have been either totally or at least extensively renovated for new, non-industrial uses. The most dramatic instances of such difficulties have occurred in the center of Tampere (which is also known as the Finnish Manchester), where a cotton mill site from the 1830s (Finlayson) and a somewhat more recent, but historically significant machinery workshop site (Tampella) are being vacated. In connection with this type of development, it has rarely been possible to achieve concrete results in protecting the historical milieu. The Finnish law on the protection of cultural monuments includes industrial monuments, but conflicts between economic and conservation interests have made it difficult to apply its stipulations.

Recent Years

In the 1980s in Finland the pace of change throughout industry accelerated. Many traditional industrial facilities have either closed down, changed hands, or disposed of their old properties, especially in urban centers. It has often been difficult to find new uses for old factory buildings and large industrial sites that have been left vacant.

There are also many encouraging instances of successful reuses, in both the private and public sectors. One of the most interesting is the Helsinki cable factory, where the empty premises were let to more than 200 craftsmen, photographers, artists, dancers, boxers, motor mechanics, art gallery proprietors, and philosophers. Within a few months, there developed, quite spontaneously, an activity center resembling Andy Warhol's well known Fac-

Fagervik: The new roof construction of the bar-iron forge was erected, and the forge restored in 1986–1987 by grants from the National Board of Antiquities. Architect Merja Härö.

tory. Today the city authorities are pondering the reuse of the cable factory, but it is hard to imagine a better content for an old factory building in a central location.

Fortunately there has also been a notable growth of interest in the industrial heritage and in old industrial environments. Plans for pulling down the old industrial quarters of Tampere, Helsinki, and Turku soon came up against opposition. The leading role played in this connection by young persons has been particularly gratifying. In Helsinki, young squatters have opposed the construction companies, and in Tampere the students have actively criticized the hush-hush agreement concluded by the city authorities with the Tampella Company.

A lot has happened in Tampere since the Nordic TICCIH meeting in 1982. At that time all the old factories on the waterfall were fully operative, except for the textile mill. In the past few years, all the industrial companies have announced plans for moving out of the city center within the next five or ten years. The Finlayson cotton mill was the first to organize an architects' competition for its precinct. The winning entry retained most of the old factory buildings but also included the construction of new apartment buildings. The winner is now hard at work on the area development plan, and there is continuing discussion of the principles of preservation to be applied at Finlayson's. At the same time the state of the so-called Old Factory is being investigated, to decide the feasibility of establishing the Museum of Labor in this building, that dates from 1837 and occupies a special position in Finnish industrial architecture.

Opposite Finalyson's, on the other side of the waterfall, a different solution was adopted. During the 1989 summer holidays, a proposal was put before the city council that implied intensive building rights for the Tampella factory district. Underlying the proposal was a planning agreement concluded by city representatives and the company with the bank and the construction company as spokesmen. The construction company actually commissioned from an eminent professor of architecture the model in which he presented a plan for four 40-story apartment blocks, a plan that would be perfectly legitimate given the building rights.This model may have helped the people of Tampere understand how the historic townscape would be transformed by this scheme. In 1990 an open competition was held for the planning of a 26-hectare factory site, the brief being to fill it up with 420,000 sq. m. of dwelling space. Although the winning entry did not measure up to the most Utopian expectations, its realization would radically transform the urban scene. As a result of the profound recession, however, there is likely to be plenty of time for studying a variety of other planning alternatives.

The situation in Tampere has shown that planning cannot proceed on a hand-to-mouth basis. Now that the revision of the Building Act has augmented the powers and responsibilities of municipal authorities, they should also have an active land policy. The revised legislation also emphasizes openness and democratic procedures in the planning process.

In Karkkila a project has been inaugurated to find new users for the old foundry buildings in the center of town. This project was planned jointly by the private company and the town authorities. Efforts have also been made to find new workplaces. The former engineering works is to be converted for use by town authorities, and new buildings—a fairly large apartment block and a swimming pool—are also to be erected within the factory perimeter.

Inquiries and Surveys

During 1989 the Ministry for the Environment published a survey report on industrial environments of outstanding historical interest in Finland. The inventory appended to that report lists a total of some 200 industrial environments. Efforts have been made to create an overall picture of the abundant tradition of Finnish industrial architecture and its preservation. One of the aims of the inventory was to investigate measures taken at various planning levels to ensure the preservation of industrial environments of historical interest. The Conservation of Buildings Act, which came into force in 1985, has made the municipalities responsible for a growing proportion of the conservation of the architectural heritage. Good examples already exist of plans for the renewal of building for the Sunila and Littoinen factory and housing areas, demonstrating the compatibility of industrial and conservation interests.

The Tampere Institute of Technology has published a major survey of the development of the secondary use of industrial buildings. That study was funded by the city of Tampere, the Ministry for the Environment, and the National Housing Board.

At the end of 1989, a report was presented to the Ministry of Education by the Maritime Museum Commission, listing 145 veteran ships and recommending that a national maritime foundation be established to finance, partly or wholly, the reinstatement of these historically interesting vessels to serve as museum ships. The commission also recommended compiling a register of veteran ships, with inclusion in the register as a precondition for the shipowner to receive financial or other

Helsinki: the industrial cityscape of Helsinki is in danger of changing dramatically. *Photo:* Johnny Korkman, Helsinki City Museum.

Pori: the machine shops dating from the turn of the century are still operational and are protected by law.

Photo: Lauri Putkunen.

support. The commission went on to emphasize the importance of old shipyards as repair centers for historic vessels.

Industrial inventories are in progress at the city and county museums of Helsinki, Turku, Lahti, Vaasa, and Pori. The last two have also published interim reports on their inventories, while Turku has published a report on the reuse of industrial buildings.

Restoration

Leineperi ironworks is located in Satakunta in the small community of Kullaa. Founded in 1771, it is beyond doubt one of the best preserved ironworking environments anywhere in Finland. It includes, for example, an old timber-clad blast furnace, a hammer forge, the manager's residence, about ten small homes for workers along the factory street, and many outbuildings. Recently both the blast furnace and the forge have become seriously dilapidated, and restoration work began, at the eleventh hour, in 1988, aided by employment subsidies. As work proceeded, it became clear that the entire framework of the blast furnace was decayed and would have to be renewed. The timber structure was raised clear with a steel support and the decayed logs replaced. Restoration work on the hammer forge is still in progress; meantime, an effort is being made to find new activities for the area. The community, with its 1,800 residents, cannot finance the restoration and reconstruction, so funds for the project have been allocated by the National Board of Museums, the Ministry for the Environment, and the County Administrative Board.

The museums board has begun restoring the system of canals at Möhkö ironworks in northern Karelia in eastern Finland. Möhkö, founded in 1849, was one of the 50 or so bog ore ironworks established in central and eastern Finland between 1807 and 1879. Its blast furnace is totally decayed, but a shelter is to be constructed to protect the ruins. The canal is to be reinstated for use by leisure craft. The three lowest locks are to be reconstructed and built of timber. In the main building, Ilomantsi-scura, the local history association, has opened an exhibition describing Möhkö's industrial history.

Meetings and Conferences

In May 1989 a meeting was held in Tammisaari, West Nyland, and Eura, Satakunta, to discuss current restoration of development projects for old ironworks in western Finland.

The Industrial Heritage Association arranged, in Kotka, in the autumn of 1989, a meeting on the subject of "Industrial Environments in Transition," dealing with the future of old industrial centers. That meeting was coarranged by the Kymenlaakso Provincial Mu-

Forssa: the cotton-spinning factory is now owned by the city and is being reused in various ways. The old factory buildings house museums, a library, an institute, and a bowling alley.

Photo: Lauri Putkonen.

Turku: the old workshop is the first production building preserved by the new Building Protection Law, which came into operation in 1985. *Photo:* Lauri Putkonen.

seum Association, the Finnish Federation of Local History Societies, the Ministry of the Environment, and the museums board. A search was made for ways to preserve and develop old industrial environments as an essential part of Finland's cultural heritage. The aim was to convene representatives of industry, national and municipal planning authorities, planners and researchers, together with the active members of the local history movement. A number of Nordic participants and lecturers were also present. Francis Sejersted described Norwegian contributions to Finland's forest industry, and Rune Svensk talked about current industrial inventories in Sweden.

The Third Nordic Seminar on Working-Class Culture took place in Tampere in September 1989. It was organized by several Finnish organizations, and its theme was "Everyday Life, Awareness and Class." The proceedings of the meeting were published in 1991.

Preservation of the bog ore ironworks of eastern Finland was the objective of meeting in Sonkajärvi, North Savo, in mid-March 1990. The participants addressed an appeal to the state, together with municipal authorities

Normarkku: the sawmill was built in the 1870s and except for its waterwheel, it has uniquely survived along with its equipment. It has been maintained by a private enterprise. *Photo:* Lauri Putkonen.

Panelia: the small dairies have declined in the Finnish countryside. The village dairy in Panelia will be used as a common cultural center and museum.
Photo: Lauri Putkonen.

and private property owners, to cooperate in restoring the few surviving bog ore ironworks.

New Industrial Museums

Finland's first foundry museum opened in 1989 in Karkkila. It is located in the middle of a foundry plant that is still active. The exhibition occupies the old head office, and right outside the door, one can see both the old blast furnace and the new foundry. The museum also includes a workers' museum with three old working-class homes. Within two years the total number of buildings will be raised to nine.

The Billnäs hydropower station at Pohja, opened to the public in 1989, was built in 1921 and is well preserved, still retaining its three original generators. This working museum has an active, central role as part of the old industrial site. The power station is owned by Oy Imatran Volma.

The Finnish Road Museum was founded in 1988 by the National Road and Civil Engineering Board. It comprises more than 60 old roads and bridges that have been preserved and signposted for travellers. Until last spring it was still intended to establish a permanent museum in Raisio, in southwest Finland, but the Ministry of Transport nonetheless decided to establish the museum in Kangasala, in Tavastland, where there is already a large motor museum.

A dairy museum opened in Saukkola in 1989. This is housed in the old granite rubble cooperative dairy, dating from 1904. At the same time, a mining museum was opened at Tytyri limestone mine near Saukkola in Lohja. Here one has the chance of an unforgettable museum visit 110 meters below ground.

The Ironworks of Pohja, at Billnäs and Fiskars

Lauri Putkonen

The Pohja district is situated in southwest Finland, close to the coast. This is in the central part of the traditional Swedish/Finnish regions, and the economical and technical influence dates from the time when Finland was part Swedish and Western culture, that is, before 1809, when Russia annexed Finland.

In the small municipality of Pohja, there are four old ironworks: Fiskars, Billnäs, Antskog, and Åminnefors. The first three were established in the 17th century and the last in 1872. In Antskog there was also one of the earliest broadcloth factories in Finland. Billnäs was formerly well known for its oak joinery products, and cutlery is still manufactured in Fiskars.

Although Pohja is still a relatively industrialized city, the decline in industrial jobs has been remarkable during the 1980s. In 1984 the Finnish Government decided to preserve the Billnäs and Fiskars ironworks. At the same time, the municipality of Pohja established a real estate company, the Pohja Ironworks Industry. It is expected that all the buildings at Billnäs and Fiskars will have been repaired by 1995.

Fiskars and Billnäs belong to the best-preserved ironworks in Finland. The oldest buildings are from the 18th century. The oldest Finnish engineering workshop was built in 1837 in Fiskars. The most representative view is that of the neo-classical manor house and the school house of Fiskars. There are also many workers' houses from the 18th, 19th, and 20th centuries. These buildings have now been renovated and will be rented to new tenants.

Fiskars: an overview.

Billnäs: the ironworks with the old water mill from the 1770s and the hydroelectric power plant from the 1920s, which is used as a mill museum. In the background old workers houses.
Photo: Lauri Putkonen.

Fiskars: stables, school, and belfry from the 1820s and 1830s.
Photo: The National Board of Antiquities.

The Industrial Center of Tampere

Lauri Putkonen

Tampere is situated in the middle part of Finland, some 20 miles from Åbo and the southwest coast It is the most industrial of Finland's cities. The earliest mills were erected on the rapids of the river Tammerkoski in the last decades of the 18th century. In the 1820s, a Scotsman, James Finlayson, established here a cotton mill that soon developed into the most remarkable textile mill in Finland. In 1837 a six-story mill with iron columns was erected. This mill is well preserved and will be be partly converted for use by the Central Museum of Labor in the near future.

In addition to the Finlayson mills there are three other fairly large textile mills: Tampella Oy, Liljeroos Oy and Tampereen Verkatehdas Oy. The latter was largely demolished in the 1970s. The first mechanical paper mill in Finland, the Frenckell Paper Mill, is well preserved but no longer used as a paper mill. In the center of the city there is also a big cardboard mill, the Tako Oy, which is still operating. Along the rapids there are hydroelectric power stations from the 1920s and 1930s.

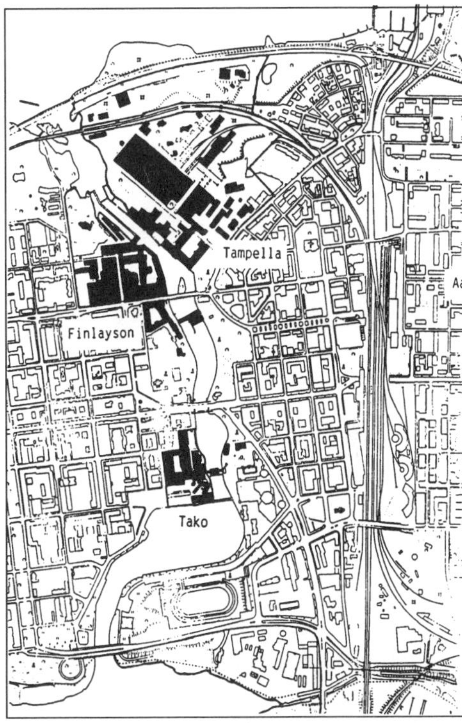

Tampere: the industrial center.

Tampere: the Finlayson Cotton Mill and Tampella Linnen Mill on the opposite sides of the Tampere rapids will soon move to new premises, and the old factory areas will be converted to new uses. The old industrial town space will, however, mainly be preserved.

Photo: the National Board of Antiquities.

Tampere: the Finlayson Cotton Mill is the oldest still-operating textile mill in Finland. In a few years, the old enclosed factory area will be reopened for use as business, offices, and accommodations. The Old Factory from 1837 will be restored as the Central Museum of Labor. *Photo:* Lauri Putkonen.

In recent years many factories have contracted or moved from the center to modern new premises. New uses will have to be found for roughly one million square meters of factory spacc. Thc city of Tampere has accomplished, in cooperation with the Tampere Institute of Technology, a large-scale investigation into the reuse of old industrial building. Two architectural plan competitions have been organized, and according to their results, there will be rebuilding—also new housing—on both banks of the river. At the same time, the city authorities of Tampere are drawing up a new general plan for the old industrial center.

The Suomenlinna Dockyard, Helsinki

Lauri Putkonen

The Isles of Suomenlinna (Sveaborg) in Helsinki were strongly fortified in the second half of the 18th century. The construction of the dry dock began in 1749 and took ten years to complete. The basins were emptied by a windmill. Later in the 1750s, this was replaced by a pump house. Under the pump vaults, two shafts were bored into the rock, and water was conducted into these from the galley basin through a vaulted tunnel. The water in the shafts could be kept running until late autumn with the help of heating stoves.

Russian rule meant a long quiet period for the Suomenlinna dock. Towards the end of the First World War, the Russians began to renew the dockyard. The walls of the basin were paved with granite cubes, and a new workshop was built with a concrete frame.

The Soumenlinna Dockyard: the workshop was erected by the Russians during World War I. *Photo:* Johnny Korkman, Helsinki City Museum.

The Soumenlinna Dockyard: the floating gate between two basins.
Photo: Johnny Korkman, Helsinki City Museum.

When Finland became independent in 1917, the dockyard was finished, and in the 1920s the State Aircraft Constructing Company ran the dockyard. At the beginning of the 1930s, the dock was taken over by the State Shipyard Company; later Valmet Oy, took over the dockyard, and the construction of the outer basin was finished. Valmet Oy continued to run the dockyard until the early 1980s, and in recent years the dockyard has been used as a repair dock for old, mainly wooden ships.

The old machinery in the workshop has been kept mainly as it was when the dockyard closed down, and the aim has been to preserve the workshop as a working museum. However, the dockyard today is still a living, real workplace in the center of the historic fortifications.

The Verla Cardboard Mill

Lauri Putkonen

The Verla Cardboard Mill is situated in Jaala in southeastern Finland, in a small, sparsely populated municipality close to the industrial center of Kuusankoski some ten miles north of Koska. It is a beautiful lake district with clear waters and historical traditions. Just a hundred meters from the mill, you can find a neolithic rock painting.

The Verla Cardboard Mill was established in 1872 as a groundwood pulp mill. Ten years later, an Austrian paper-maker, Gottlieb Kreidl, bought the factory and rebuilt it into a groundwood pulp and board mill. In the 1890s brick replaced timber in the mill buildings. The buildings, which represent a rather decorative brick architecture, were designed by Edvard Dippel.

The Verla Cardboard Mill: the brick buildings by the power canal date from the 1890s.
Photo: Lauri Putkonen.

In 1922 the Verla Mill passed into the ownership of a large company, Kymmene Corporation, and in 1964 the mill was closed. Kymmene Corporation transformed the site into a recreation center for its employees. In 1970–1971 the old mill was carefully restored to become an authentic mill museum, and it is undoubtedly the most remarkable industrial museum in Finland. The interior of the mill is unchanged, with several of its ancient machines from the turn of the century left, and the atmosphere of the old workplace is easy to sense. The owner's residence was converted into a restaurant. During its last days, when the mill was still operating, the whole process was filmed, and this film is now shown to visitors.

The Verla Cardboard Mill: the factory halls have load-bearing walls with intermediate floors supported by wooden pillars and beams.
Photo: Lauri Putkonen.

The Sunila Pulp Mill, Kotka

Lauri Putkonen

The Sunila Pulp Mill: the brick factory buildings in their original form.

Photo: Roos, Museum of Finnish Architecture.

The Sunila Pulp Mill was erected mainly in 1937–1938 to produce cellulose sulphate and was located on the coast outside Kotka on the Finnish Bay of the Baltic Sea. The industrial plant, with its large village for employees, was designed by Alvar Aalto.The village, begun after the mill was nearly finished, has housing for all categories of plant staff. The rows of apartments are scattered along the wooded hillside. In Finland Sunila is considered one of the most remarkable monuments of functionalism.

The Sunila Pulp Mill is still operating, and there have naturally been some structural alterations in the mill building. In the new plan of 1987, the residential area is protected and there is a recommendation that the general appearance of the industrial plant be preserved.

The Sunila Pulp Mill with its large village for employees was designed by the architect Alvar Aalto and built mainly in 1936–1939. Workers' and employee's row houses integrated into the surrounding environment still represent a high standard of living.

Photo: Kari Hakli, Museum of Finnish Architecture.

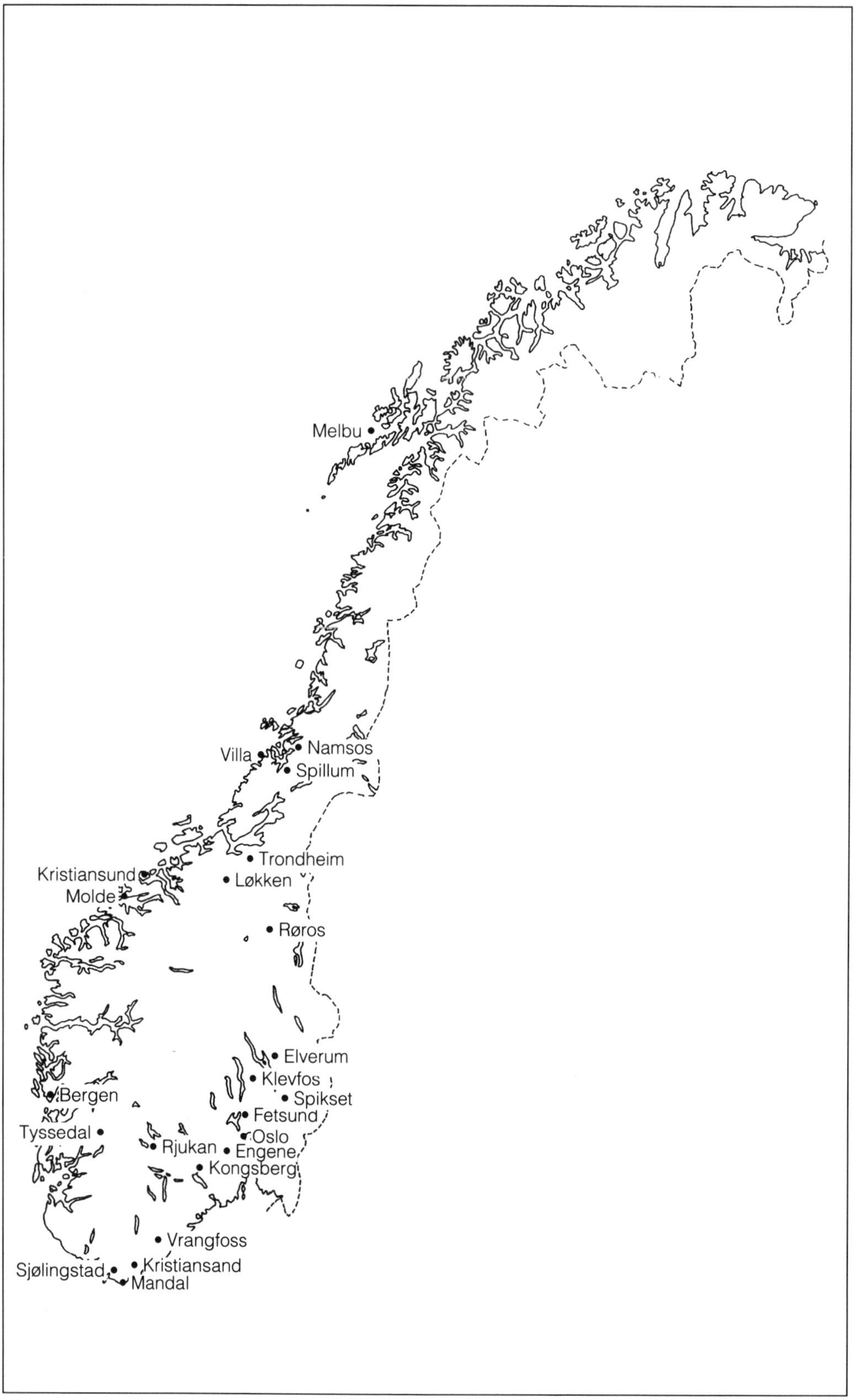
Melbu
Villa
Namsos
Spillum
Trondheim
Kristiansund
Løkken
Molde
Røros
Elverum
Klevfos
Spikset
Bergen
Fetsund
Tyssedal
Oslo
Rjukan
Engene
Kongsberg
Vrangfoss
Sjølingstad
Kristiansand
Mandal

Norway

Ola H. Øverås

The View of Conservation Changes . . . .

"In Røros, we are kindly informed . . . there is an old building which is of very singular appearance and apparently deserves to be drawn." This rather sober report in the 1891 yearbook of the Society for the Preservation of Norwegian Ancient Monuments refers to a single building from 1646 in a mining village, not quite 150 km southeast of Trondheim, which for the past 13 years has been included in the UNESCO "World Heritage List."

Like so many other things in life, views about what should be preserved mature and develop. That process, however, is steadily accelerating. For several years now, powerful forces have been working to preserve landmarks of industrial significance from the past 100 or 150 years, i.e., from a period long *postdating* the publication of that yearbook report. Preservation efforts have, for example, included periods of architecture that were long referred to as a "medley of styles" and that, for a few years after the Second World War, were still regarded skeptically or, at best, with lukewarm interest.

It might have seemed to private persons—and may perhaps still do so—that there were two groups of preservation campaigners, each with different approach. One group, consisting largely of architects and art historians, might seem to be mainly interested in buildings. Another group, consisting of museum people and others more or less directly connected with industry and other technical environments, could for its part seem to be mainly preoccupied with what had been going on inside the buildings. But the antithesis was not really that drastic, and today at least, the participants display a more balanced view, and, as a rule, the two groups cooperate.

. . . and So Does the View of Museums

Typically, the building said to be worth drawing is no longer to be seen in Røros. Early in this century, it was transferred to the district museum in Trondheim, and, equally typically, in recent years there has been an ongoing debate about moving it back to Røros again.

Over many years, the Røros copper works had assembled valuable historical collections

"The Industrial Fairy Tale," by Nils Berglien (1853–1928). The painting now hangs in Tyssedal, which along with Rjukan, Eykehavn, Sauda, Odda Høyanger, and Årdal, is one of the most interesting hydroelectrically powered industrial centers of early 20th-century Norway. Together with several mining centers, these places emerged rapidly from nearly nothing into rather well-planned and designed villages and towns.

Source: Bj. Eidnes.

in a fine old industrial building. A proper Røros museum did not materialize until 1930, but the location outside the town turned out to be a "back-water". After the copper work went bankrupt in 1977, its main buildings, among other things, were taken over by the museum, and plans could now be made for the museum as a living part of the environment in which all aspects of the history of the mining community and its surroundings could be illustrated by walks from the center to outfields and mountains.

The Kongsberg silver work acquired a museum of its own in 1938, but because of the war, this was not opened, in the old furnace dating from 1844, until 1945. This again has early collections, partly assembled in connection with an exhibition in Paris in 1878. Those collections have now begun to be ascribed a historical value in their own right, after long being looked on as ghosts from a stylistically dubious period.

In practice, the Silver Work Museum became, in 1965, a central museum of Norwegian mining and manufacturing, and this was confirmed in 1983 by the addition of "Norwegian Mining Museum" to its name. Owing, however, to lack of funding and hidebound regulations, it was inadequately equipped for its task. A recent reorganization has put this right, and both the Silver Work Museum and the Museum of Mining have begun to flourish in a manner that inspires great expectations about the many activities occurring there.

Old Mining Industry Popular

We have dwelt at some length on mine workings, because these have traditionally dominated the preservation of industrial landmarks, being both numerous and distributed all over the country. Then again, they are frequently romantic, despite, or perhaps indeed because of, the rigorous life they represented, often with an aura of mystique and never failing to stir the imagination. The first industrial landmark in Norway to come to statutory protection, in 1959, was in fact a shaft furnace from 1867, in Nåverdalen, in the far north of Hedmark County.

Diversity

To complete the picture and to convey something of its abundance and diversity, we can list examples from the history of technique, buildings, constructions, and environments in addition to the mining industry that have been preserved or whose preservation is being seriously considered.

- The sunken roads or "holloways" round Oslo Fjord, or Northmen's track over the Hardanger plain, all of which date back to earliest times.
- The bog ore industry of Hardanger plain or Nord-Trøndelag, which at all events was operating on a large scale at about the beginning of the Christian era.
- The soapstone workings of Kvikne in Nord-Østerdalen, from the 4th–5th centuries B.C., or the millstone quarries of Rønneset in Outer Sogn, which were probably being worked c. 500 A.D.
- The silver road from Kongsberg, our first highway built by royal command, in the 1620s, or again the main road between Southern and Central Norway via Dovre, which was first used by a carriage in 1704—true enough a royal carriage and with a great retinue of people and horses.
- The house for children, correction, and manufactory in Bergen, built soon after 1702, or the Trondheim sugar factory from 1754.

The Red Bridge at Røros. The Aspaas Manor was erected shortly before 1800. In 1917, it was moved to the Trøndelag Regional Museum. At present plans are underway to return it to its original site.

Akerselva on the river Aker is the birthplace of Norwegian industry. The river is about ten kilometers long with a falls area of approximately 150 meters divided into 22 separate falls. The water power has been exploited since the Middle Ages. To the right is the oldest remaining industrial building in Norway, the Glads Grain Mill. It was originally built as a paper mill in 1732. The adjacent building houses the Hjula Weavery built in 1855.

From a Xylograph.

- The timber-driving canals at the north end of Femunden, from the 1720s, or "Kjerraten i Åsa," an installation from 1807, northwest of Oslo, which pulled timber from a "foreign" watercourse up to the proprietor's own, which was about 400 m. higher.
- A beacon in Alvdal, Nord-Østerdalen, built in 1812, or the Villa Lighthouse on the coast of Nord-Trøndelag, erected in 1839.
- Norway's first cast iron bridge, Løkke Bridge from 1829 at Sandvika, near Oslo, or the first chain suspension bridge, Bakke Bridge from 1844, over the River Sira in Vest-Agder.
- The Stekka gangsaw in Hardanger from 1823, or Harkmark Mill in Aust-Agder, dating from 1869.
- Spikset farm brick kiln in Sør Odal, or Sødal Farm lime kiln, not far from Kristiansand, both dating from about the mid-19th century.
- The Bredalsholmen dry dock at Kristiansand from 1873–1876, or the Brunsviken rope walk outside Kristiansund, founded in the 1850s.
- The Berger textile mill in Vestfold, started in 1880, or Helle brickworks in Outer Sunnfjord, from the 1890s.
- The Fetsund timber boom system chutes northeast of Oslo from 1862 or the Telemark Canals, completed in 1892.
- The Kistefoss pulp mill, northwest of Oslo built in 1899, or the Klevfoss paper mill as rebuilt in 1909.

Kongsberg. The center of the Silver Town with its famous silver works is well preserved. Most of the buildings date from the middle of the 18th century, among them the first European school of mining engineering built in 1757. The photo depicts workers' dormatories built between 1866–1874.

- The penstock of Vemork power station on Rjukan built around 1910, or Ringedalsdammen, an enormous dam of dressed granite rubble, constructed 1910–1918 as part of the first major hydropower development in Tyssedal, Hardanger.
- Skansen bascule railway bridge in Trondheim constructed in 1919, or the Norddalen pendulum bridge from 1902 on the Ofoten railway near Narvik.
- The big Neptun herring oil factory at Melbu, Vesterålen built in 1910, or the Spillum Steamsaw and Planing Mill in Namsos, dating from 1884.
- A North Sea oil rig from 1980 . . . ?!

A long list, but still including only a small portion of Norway's industrial landmarks in terms of both quantity and variety. Where many of the categories are concerned, we can only guess at the total quantity. We have yet to command an adequate conspectus.

Registration

A succession of inventories have been undertaken through the years. One or two of them were intended to cover the entire country, but most, some of them exhaustive, have been inventories of particular items or installations, or groups of the same, often in connection with post-graduate theses or main subjects, local histories, commemorative publications and, the like. These limited inventories often included solid, comprehensive descriptions, but were less predictable about serviceable maps, measured drawings, and photographs. Better coverage of this kind is often, though not always, obtained in connection with concrete preservation measures.

Nationwide inventories, understandably enough, are relatively scarce where written information and illustrations are concerned. First and foremost, they have to convey a necessary overview, so that more thorough investigations and preservation measures can be undertaken where they are most urgently needed.

The first nationwide inventory was compiled in connection with the 1920 Preservation of Buildings Act. Letters containing simple questions were sent to parish priests, high sheriffs, and other reliable persons in local communities. The result was meagre for the history of technology and industry, for despite its name, the act also applied to technical landmarks of all kinds.

The same is fundamentally true of the other three nationwide inventories. The five museums of archaeology in Oslo, Stavanger, Bergen,

The Sugar Refinery, 1754, is probably the oldest industrial enterprise in Trondheim. In 1856 it was converted into the E. C. Dahl brewery. The brewery moved to new quarters some hundred years later. It is now an office building.

Trondheim, and Tromsø have long been recording cultural monuments pre-dating 1537, but only in exceptional cases do these inventories include the remains of industrial landmarks. These are few compared with burial mounds and dwelling sites, and perhaps less easy to discover because they are not the sort of thing one expects to find. In the case of large workings such as the soapstone quarries of Kvikne or the millstone quarries of Selbu in South Trøndelag, pure chance led to the discovery of their antiquity. The reindeer-trapping installations in the high mountains are an exception to this rule. They have been more systematically recorded and have in some cases proved to be systems extending for several hundred, if not thousands of meters.

Early in the 1970s, the Department of the Environment compiled a county-by-county inventory of natural and cultural resources, but its content varied according to what the individual county could track down. Then, in the mid-1970s, the "Register of Post-Medieval Monuments," popularly known as SEFRAK got started in earnest. The cost of the inventory was shared equally between the state and local authorities. In principle, this inventory includes all permanent cultural remains post-1536 and pre-1900, but pre-1945 in Finnmark and other regions that were ravaged during the Second World War. In practice it is buildings above all that have been recorded, with little if any information about their contents, insofar as they are industrial buildings. A separate form for other permanent cultural remains than buildings has also come to be used slightly more extensively with the passing of time, and this ought reason-

Villa Lighthouse, Flatanger, west of Namos in the Trondheim fjord, was built in 1839 and was one of the earliest lighthouses to be charcoal fired.

Photo: Anne Winterthun 1981.

ably to mean the inclusion of a number of technical and industrial landmarks. SEFRAK's field-work is scheduled for conclusion in 1995. Each county administration's conservation officer is responsible for local activities, including directing the field-work. There is work in progress to coordinate all heritage recording systems into a single system called NOREK.

The main foundation of many inventories of technical landmarks is the achievement of Gunnar Thuesen, for many years a curator at the Norwegian Museum of Science and Technology, who died recently. Starting in 1943, and on top of his other duties at the museum, he systematically measured, photographed, and collected information about early technical installations and the remains of such installations all over the country. For a year or two from 1969 onwards, he did so with a grant from the Central Office of Historic Monuments and Sites. Unfortunately, for financial reasons, the extent of this inventory was limited.

Løkke Cast Iron Bridge at Sandvika, 15 kilometers west of Oslo, was built in 1829, fifty years after the famous "Iron Bridge" in England. The elements were cast at the Bærums iron works. It is one of two such bridges built in Norway, the other being eight years younger and four meters shorter is at Fossveit and was manufactured at the Næs ironworks. 150 years after its construction, the Løkke bridge was moved some 80 meters north and is preserved today for pedestrians and cyclists. *Photo:* Ola Øverås 1977.

Mills

The National Grain Corporation, which has a monopoly of grain and flour imports and exclusive purchasing rights for Norwegian grain, and is one of the largest business concerns in the country by turnover, compiled a national register of water mills between 1955 and 1965, under the auspices of the Water Mills Committee. Not all mills were registered when it came to the point, but the inventory resulted in 42 of them becoming listed buildings, together with eight grain dryers and one village mill. The Grain Corporation also paid for reinstatement and concluded maintenance agreements with the owners. During the 1980s, the Grain Corporation, acting in consultation with the Norwegian Museum of Science and Technology and the Central Office of Historic Monuments and Sites, compiled an inventory of all village and district mills in the country—a total of some 1,100. More than 40 of these have been designated of outstanding interest, including 15 that have been given special priority.

Roads

At about the same time, the National Road Administration began working for a more systematic preservation of old roads and bridges, implements, machines, buildings, and equipment, as well as written and verbal information concerning the history of roads. A "watchdog" was appointed for each county road authority to prevent the destruction of historic buildings and structures, equipment, and records, and for the collection and preservation, in a proper way, of artefacts and mate-

The six locks built in 1892 at Vrangfoss on the Bandak Canal, one of the Telemark canals, are a highlight of the system. Intended originally to transport timber, the oldest part of the Telemark canal system, the locks at Skien, date from the 1850s. The other major system is that on the Halden rivers completed in 1877.

Photo: Skøien 1898.

rials relating to the history of roads in Norway. These "watchdogs" meet once a year for a briefing and to discuss problems and experiences and also, not least, to visit current preservation projects. One result of this work is the Norwegian Roads Museum, which will open to the public in the summer of 1992 at Hunderfossen in Gudbrandsdalen.

Railways

The official opening of the Roads Museum comes 96 years after the opening of the Norwegian Railway Museum, the largest of its kind in Scandinavia, which has been located in Hamar since 1955. The museum itself includes a short length of historic railway line, but all over the south of Norway, railway enthusiasts have rescued four lines of different kinds and from different periods. Two of these were industrial lines and are electrically powered. The longest runs between the former Løkken Verk southwest of Trondheim, and what used to be its melting plant and export harbor on the fjord, a distance of almost 26 km. Today only 19 km are in use. This is also the oldest electrified railway line in the country, completed in 1910.

When, at about the same time, Norsk Hydro began producing artificial fertilizer at Rjukan, they did so because this was where they produced their electrical power. As yet, power could not be transmitted over long distances without many hazards and heavy losses. Dependable lines of communication therefore had to be built to convey materials and machinery to and finished products from the factory. A railway operated from Rjukan, originally all the way from the Vemork power station, and down to the ferry port of Mel on

The Spillum Steam Sawmill built at Namos in 1884 is a well-preserved example of the 200 some steam powered mills built in Norway since the 1860s.
Photo: Jan Wiig 1989.

Tinnsjøn. That is still the home port of the Ammonia, the world's oldest rail ferry and perhaps the only steam rail ferry still in working order.

Veteran Circuit

Looking slightly further ahead, the train from Rjukan and the Tinnsjøen ferry are to be included in an exciting round tour of the center of southern Norway, a tour that can come to include a trip on the Telemark Canal network, changing to antique buses and cars or an old-time horse-drawn carriage for the return journey to Rjukan. This journey will pass through scenery ranging from the wild and rugged to the lush and fertile, passing by stave churches, medieval industrial landmarks, centuries-old homesteads, and National Romantic tourist hotels from the second half of the 19th century. There will also be opportunities for trips in "antique cable cars" or on foot or horseback along dramatic bridleways and packhorse trails.

Watercourses

When operations connected with a watercourse are discontinued, the law requires someone to tidy up. That someone may be the proprietor or operator, but in the absence of any such identi-

The Offerdal Sawmill (c.1860) and grainmill (c. 1890) represent one of the few preserved small-scale local mills.
Photo: Ola Øverås 1980.

The Wool Factory at Sjølingstad, a few kilometers north of Mandal in West Agder, is the only complete wool factory in Norway today. Founded in 1894, it has been in continuous production for 90 years. The building dates from 1912.

Photo: Geir Nerhus 1990.

fiable party with financial capacity, NVE, the National Energy Board, steps in. During recent years, however, the board has become increasingly aware that much of what the law expects to be removed is really of great historical interest. Instead of demolishing buildings, channels, and dams, perhaps the money allocated can be applied to their reinstatement. Accordingly, the board has compiled recommendations on the "treatment of historic riparian installations," firmly advocating more generous funding for reinstatement and preservation and a good deal more care and consideration before making the final choice between demolition and reinstatement.

Parallel to the work of this committee, the Hedmark county authorities have compiled an inventory of all buildings, installations, and stray finds connected with log driving, and traces of the same. These inventories have been made the basis for the prioritization of individual features and environments for preservation in the county.

National Council for Cultural Affairs

The state of activity we are now witnessing for the registration and preservation of the industrial and technical heritage is very much due to an initiative taken by the Norwegian Council for Cultural Affairs. In 1974 the council, prompted by a steadily growing number of applications, both to the council itself and to the Central Office of Monuments and Sites, for grants for the reinstatement and preservation of landmarks of industrial significance, appointed a committee on the preservation of the technical and industrial heritage. In addition to finan-

The Brunsvikens Ropewalk was built in Kristainsund in the 1850s and is now preserved in the local museum. It still produces specialty ropes for Mellomverven, a repair shop for vintage ships.

Photo: Jan Wiig 1978.

The Old Sild Oil Factory (c. 1910) in Melbu has been purchased by the Central Board of National Antiquities, in cooperation with Melbu Fish Companies and the local museum, to be preserved as a national museum for the fishing industry.

Photo from 1985.

cially supporting various specific cultural projects, the council has the express objective of inaugurating new schemes that it finds necessary in various cultural sectors. The Register of Post-Medieval monuments in Norway was inaugurated by the council ten years earlier and followed, a little later, by an action for the more systematic conservation of ships.

Report on Preservation

In 1988 the committee thus appointed by the council, consisting of representatives of the Norwegian Trade Union Federation, the Federation of Norwegian Trade and Industries, the Norwegian Council for Cultural Affairs, the Norwegian Museum of Science and Technology and the Central Office of Monuments and Sites, presented a report that, in addition to proposals for the organization of preservation activities, included valuations and priorities based on more-or-less comprehensive reviews of the situation in the various branches of economic activity. As work on the report progressed, it became a source of inspiration, and parts of the committee, aided by new forces, were commissioned to follow up ideas presented in the report, in collaboration with local interests and with representatives of organizations in the employment sector, county authorities and relevant authorities at the national level. The various national institutions are increasingly acknowledging their duty to preserve both visible and invisible memorials of developments in their fields, as, for example, in the case of the National Road Administration.

Statutory responsibility for the technical and industrial heritage has been vested in the Central Office of Monuments and Sites since 1920, when the first legislation was passed concerning the secular cultural heritage. At the New Year 1991, the Central Office of Monuments and Sites also took over the main burden of financial responsibility from the National Council for Cultural Affairs, but on a basis of continuing co-operation with the council, with relatively clear lines of demarcation between the two authorities. The Central Office of Monuments and Sites awards grants for reinstatement, restoration, and, occasionally, reconstruction or, very occasionally, the purchase of buildings, installations, equipment, machinery, and plant that have remained in, or can be returned to, their original location. After a recent reorganization, each county administration within the cultural sector has been allocated a very central role in this work. The council for its part supports all manner of efforts for the collection of facts about the cultural heritage, the preparation of that heritage for museum use and its historical presentation. Many museums of a more traditional kind have, over the years, contributed generously towards the conservation of industrial monuments in situ, frequently with reference to an ecological context.

With so many projects in existence and so many new ones being added all the time, an overview is liable to be out of date as soon as it is completed. However, at the time when the main responsibility was taken over by the Central Office of Monuments and Sites, there were 35 publications on the preservation of single features or groups of features and 27 more were in preparation.

National Government Priorities

The Ministry for Environment Protection, to which the Central Office of Monuments and Sites reports, has given priority to the preservation of memorials of the working and leisure life of ordinary people, with special emphasis on the situation of women. In addition, priority is to be given to coastal culture. The Neptun herring oil factory and fish processing museum in Melbu, Vesterålen, is a good example of a feature amply qualifying for inclusion in all three of these priority fields. The Ministry has also instructed the Central Office of Monuments and Sites to put forward a draft national plan for landmarks of technological significance before 1993. That plan is to be based on the report submitted in 1988 by the committee of the Council for Cultural Affairs, but much work still remains to be done since, inevitably, the committee's report was more detailed and exhaustive in some fields than in others.

Coordination of Efforts

The Central Office of Monuments and Sites has appointed one extra person to work on the national plan. In addition, a new joint group has been set up, comprising representatives of the Norwegian Confederation of Trade Unions, the Federation of Norwegian Industries, the

Blåfarveverket, The Cobalt (Blue Color) Works at Modum.
Photo: Werenskjold 1897.

Norwegian Museums of Art and Cultural History, the Norwegian National Council for Cultural Affairs, and the Central Office of Monuments and Sites. Some "old sweats" from the Norwegian Museum of Technology, the National Council for Cultural Affairs, and the Office of the Central Office of Monuments and Sites are also taking part, as observers, to ensure continuity. This joint group is intended to act as an advisory, inspirational, and facilitating agency for those charged with putting the projects into effect.

The Norwegian Museums of Art and Cultural History, an umbrella organization representing most of Norway's museums and their employees, recently set up a special museum group for industry, communications, and installations to organize more active efforts in these fields by parts of museums. It is also to be hoped that this group may act as a national organization of TICCIH.

It is perhaps appropriate that this overview should end where it began, with the mining industry. Widespread wishes for the preservation and care of mines, smelting plants, and all the memorials of the mining industry are colliding more and more often with a fear of adverse occurrences at the features preserved. There is a fear of accidents at inadequately safeguarded installations, but above all, of the risk of heavy metals escaping from malm ore and slag heaps and from settling ponds and flooded shafts.

The Ministry of Environment Protection has also appointed a committee on this subject, to evaluate the various considerations in relation to one another and, if possible, to arrive at a list of preservation priorities based on as complete a picture of the situation as possible. This committee is made up of representatives of the State Pollution Control Authority, the Commissioner of Mines, and the Central Office of Historic Monuments and Sites and, of necessity, is collaborating closely with the Norwegian Museum of Science and Technology and, not least, the Norwegian Museum of Mining, both of which have done sterling work in compiling a proper overview of the country's almost innumerable mining installations.

Conclusion

On the whole it is fair to say that the preservation of the landmarks of technology and industry in Norway has the wind in its sails at present and that time is short if equilibrium is to be established in the overall conservation of the country's cultural heritage. Disappointments, however, are bound to happen. A brickworks at Fredrikstad, a distillery in Hadeland, and a fish-processing plant in Finnmark are examples of ideal preservation projects that have slipped through our fingers in recent years.

Whatever the reason—evil intent, ignorance, bad luck, or simply private considerations and other social interests being more important—all that remains is for us to learn as much as possible about such events and to convert disappointment into inspiration. In view of the developments that this essay began by describing, we can at worst console ourselves—half jokingly, half seriously—with the thought that "most of the cultural heritage has yet to be created."

On the other hand, we will definitely oppose the "creation" of some parts of that heritage, only to campaign later on for their preservation. This is not necessarily as inconsistent as it may seem. Conflicts connected with a proposed construction can, once it has nonetheless materialized, invest it later on with an extra degree of historical interest. But anyone taking this as a plea for the wholesale award of permits has completely missed the point!

The Mining Town of Røros

Ola H. Øverås

Mining operations at Røros are connected with the first major expansion of Norwegian industry in the 17th century. King Christian IV pinned great hopes on the potential of Norwegian mining—copper mining in this case.

Operations eventually started in 1644, and the mining town was founded two years later close to Åsen Farm, which has changed little in the past 150 years and has remained within the same family. The entire village, except for the church, was burned to the ground in the wars of 1678 and 1679, but comparing the situation today with the two oldest known maps, from 1658 and 1711 respectively, one can see that the layout of the mining town resulted from two main influences. The more immediate of the two, the larger town of Trondheim, which also provided much of the capital, remained a medieval town until the fire which swept through it in 1681. And on the other side, Kristiansand on the south coast was founded in 1641 by royal command and with the gridiron plan typical of the age.

The first church at Røros, built of timber, was replaced in 1784 by a stone edifice with seating for 1,540. Otherwise most of the existing settlement dates from the past 200 years and can still be characterized as a collection of closely huddled farms. Farming was a common sideline among both workers and overseers at the Røros copper works. This was an advantage to the works—relieving as it did some of the problems of obligatory food supply—but the main advantage rested with the workers. The mines are a good way north of the

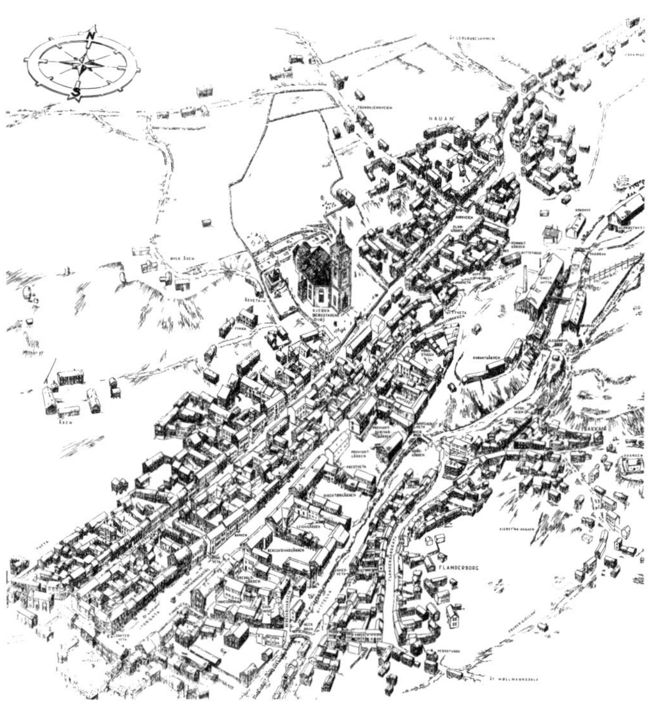

The mining town of Røros in the 1890s. The smelting house is by the river; opposite are the coal house and slag pile. Close to the smelting house are workshops and offices. As a whole the town is very little changed today, apart from the modern industrial buildings and dwellings in surrounding areas.

Reconstruction and drawing in 1944, Arne Berg.

Røros c. 1715. B–smelting house; C–walls for calcination;D–"cold calcination"; E–ramps for charcoal; F–the sawmill; G–the Hut River.

Source: The Royal Library, Copenhagen.

town, which was sited to provide adequate water power for the furnace. Situated 650–900 meters above sea level and afflicted during winter by constant blizzards and temperatures as low as minus 50°C, all the heavy transport operations, which were absolutely dependent on winter ground conditions, must have involved a degree of privation and exertion that today is barely imaginable.

As a result of the immense consumption of timber, both for building and—above all—for firing and charcoaling, furnaces gradually came to be built closer to the available forest. It was cheaper to carry a load of copper ore to the charcoal pit than to carry three or four loads of charcoal to the copper ore. This also helped to bring about the two-way interaction generating a distinctive culture within a wide radius around the mining town. In addition to a solid agrarian culture, the mining environment was also surrounded and doubtless influenced by Lapp (Saami) culture.

Many clear traces of the activities of the copper industry around about Røros are still extant, e.g., stone-filled timber dams for regulating the water supply and channels for floating down timber from remote forests, or again the employees' living quarters and the managers' country retreats.

The Røros Museum, founded in 1930, gradually encompassed several buildings outside the town, but by 1930 a number of buildings in the town center had been listed for ten years and there was a steady growth of interest in the actual mining town as a historic monument. In the 1970s the museum began making plans to transfer to the center of town. When the copper mill went bankrupt in 1977, after 333 years' continuous operation, and most of the business was taken over by the state, including the parts of historic interest (which were not really all that old), the museum was able to beef up its plans. The last smelting works, like its numerous predecessors, had its interior destroyed by fire, but important parts of it were still standing. It was decided to reconstruct the smelting works, concentrating mainly on the exterior but also including that of the interior, which was still intact, and also to provide the museum with offices, stores, and exhibition facilities.

The furnace of the Røros Copper Works, 1889.

Olav's Mine, the workings of which were in use between 1936 and 1973, is now used as a demonstration mine. It is administered by AOF, the Workers' Education Association, which also runs a camp school in another mining area not far away. To reach down to Olav's Mine, one has to pass Nyberget Mine, which was opened as early as 1650. This provides an excellent opportunity of illustrating the span of mining operations over three centuries. At the end of the 19th century, the Røros Copper Works was one of the biggest industrial concerns in Norway and open to new possibilities. One such was electricity. In 1896 the mine commissioned the Kuråsfossen power station, at that time the biggest in Scandinavia. To convey electric power to the mines, blast furnace, and elsewhere, the company constructed a 24 km power transmission line.

Røros has had an unusually large number of local newspapers. One of the most enduring and best-known was *Fjell-Ljom*. The *Fjell-Ljom* building, with its machinery and equipment, is now to be made part of a foundation together with the Røros Museum, the Olav's Mine, Kuråsfossen, and some other, smaller collections.

Since 1959 the Central Board of National Antiquities and Sites has been providing annual grants, which have been used in consultation with the museum and with the municipal authorities. Røros has now acquired a town custodian of antiquities, employed by the municipality.

Over the years, starting in 1920, a good 80 buildings have been preserved, and most of the town center has been put under a preservation order as a special area. But the mining town of Røros is not just a big museum. It remains the vigorous, flourishing town it has always been.

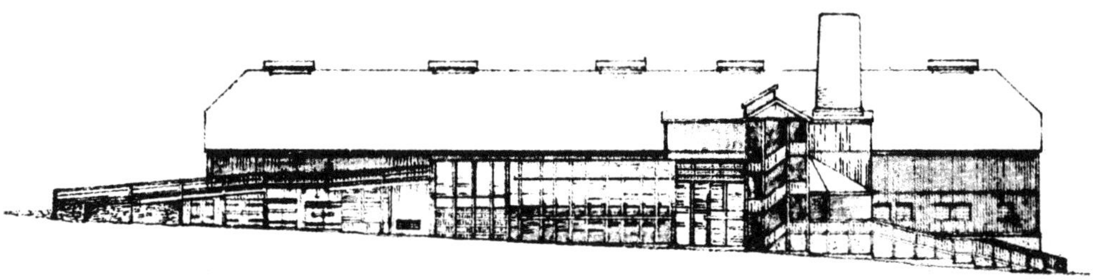

The new museum building designed by Seppo Heinonen at the Røros Copper Works.

Klevfos Cellulose & Paper Mill and the Norwegian Wood Pulp and Paper Industry

Magne Jervan

Abundant access of raw materials and hydraulic power were a good starting point for establishing a chemical and mechanical wood pulp and paper industry in Norway at the end of the 19th century. This industry was long of prime importance with regard to exports and

A drawing of the museum at the Klevfos Cellulose & Paper Mill.

employment. Klevfos Cellulose & Paper Mill was established in 1888 and was one of the pioneers producing so-called kraft paper from sulphate pulp. The factory was, however, relatively small within this line of business. The highest number of workers here was about 120, and at most 3,000 tons of chemical wood pulp were produced per year. In 1909 the factory was destroyed by fire, but it was rebuilt within two years. After that the machinery was only insignificantly improved. Thus, when the factory closed down in 1976, its buildings and machinery represented a past era in a high technology industry.

The Establishment

Shortly after Klevfos Cellulose & Paper Mill was given up, local historians and some of the former workers had the idea of preserving the old factory. This coincided with a growth of attention in industrial archaeology of antiquarian professionals.

The society known as "Friends of Klevfos"

Plan of the museum with the mill at its center.

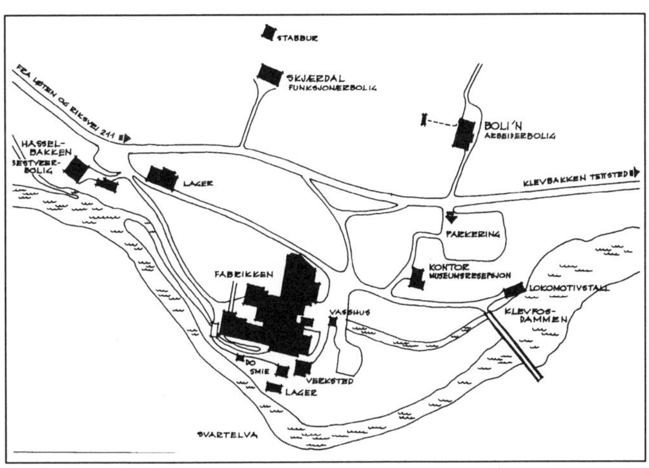

A view of the mill in 1976, just before it was closed down.

was founded in 1978. On their initiative and as a result of their efforts, restoration of the buildings began in 1980, mainly in the form of grants from national antiquarian authorities. Plans for the work were drawn up by the director general of the Central Board of National Antiquities. That same year, Klevfos Industrial Museum acquired administrative links with "Hedmarksmuseet og Domkirkeodden," an institution responsible at regional level. Until now, the economic position of Klevfos Industrial Museum has been satisfactory. The museum receives an annual grant out of public funds. Every year since 1982, this has been augmented by donations and grants from various quarters. For example, Klevfos Industrial Museum has an agreement with the local labor market authorities to employ jobless skilled workers on restoration work. The Norwegian Trade Union Confederation takes a close interest in the museum and has donated considerable amounts of money for various projects. "Friends of Klevfos" is still very active and money donated by this society has now provided the museum with a personal computer.

Klevfos Cellulose & Paper Mill in the late 1980s when the mill was converted to a museum.

The Site

Klevfos Industrial Museum consists of the old factory buildings with their complete machinery *in situ*, the dam, the canals, the garage for the steam engine truck, the mill manager's house, the house for the office manager and shift foremen, and an eight-flat tenement building for the workers and their families. Klevfos Cellulose & Paper Mill's nearly complete records are also kept at the museum. The area is about 323 acres.

One of the mill's paper-making machines from 1911.

Photo from the 1970s.

Preservation, Substantiation, and Information

A long period of minimal repairs and two years' standstill in the old factory had left it in an advanced state of decay. Restoration of machinery and buildings, therefore, has necessarily been one of the main tasks till now. Great importance, however, is attached to the human aspect as well as to technology, and great efforts have been made to document the daily lives of workers and other employees, both at work and during leisure hours. Several former workers and their families have been interviewed.

The museum was officially opened on June 13, 1986. Since then the public has been welcome to visit the whole site. In the workers' tenement building, they can also see a flat that has been fitted up, showing how a family lived in the 1930s. The furnishing of this flat is based on interviews with the previous lodgers, and much of the equipment is what they actually had when they were living here. An old married couple who had kept their furniture, linen, and so on for several years, have now donated everything to the museum.

The most important plans for the coming years are as follows. The restoration of machinery and buildings will be completed in about 1990. Apart from the history of the old factory and its surroundings, the development of the Norwegian wood pulp and paper industry and the population affected by it will be illuminated through research, publications, and exhibitions. A separate exhibition building is now being planned. This being a pioneering project, the museum management has established contacts with national and international antiquarian professionals in industrial archeology. Their experience has been a major source of guidance and inspiration in the work done so far. This, together with continuing cooperation with the local population and organizations connected with the industry, will be no less significant in the future.

The Nitroglycerine Factory at Engene

Torleif Lindtveit

The explosive nitroglycerine was invented by the Italian Ascanio Sobrero in 1847, but it was Alfred Nobel who developed the invention for industrial production, taking out a patent in 1863. He then went on working to devise a method that could make this powerful explosive safer to use. He found that kieselguhr (fuller's earth) could absorb three

The nitroglycerine factory at Engene. The museum's centerpiece is the nitroglycerine factory itself. The acid station is on the left

or four times its own weight of nitroglycerine. Thus nitroglycerine was turned into a plastic explosive. His first patent was registered in Britain in 1867, and he called the new explosive Dynamite.

Norway soon came to be involved in this new industrial development. Nobel sold his patent for the production of nitroglycerine to two friends and colleagues who moved to Norway, founding Nitroglycerin Compagniet

The laboratory as it was in 1897.

A modern equivalent of the apparatus used at the nitroglycerine factory at Engene. is constructed of stainless steel and is on display at the museum. It is considerably larger than the original and capable of nitrating 150 kgs batches on nitroglycerine.

there in June 1865. The factory was built at Lysaker, a little to the west of Oslo. Production already began in August that year, making this the second nitroglycerine factory in the world. Dynamite production also began here in January 1868, and dynamite entirely superseded nitroglycerine during the next seven years.

In 1874, however, the factory was devastated by an explosion. For safety reasons the new factory was built at Engene, Setre in Hurum about 50 km south of Oslo, on the western side of the Oslo Fjord. It was completed in 1876. This factory in turn was laid waste by an explosion in 1896, but it was rebuilt on essentially the same lines as before. It was closed in 1976, by which time it was probably the only factory in the world still based on the methodology of Nobel and his associates, as the process evolved in Nobel's lifetime, through the work of Liedbeck, Lamm, and others. This prompted discussions concerning the fate of the building. The owner, Dyno Industrier A/S, founded in 1971 though a merger of Norwegian explosives manufacturers, got in touch with the director general of the Board of National Antiquities and Sites and with the Norwegian Museum of Technology. Both were strongly in favor of the factory being preserved as an industrial landmark. The owner, together with a group of enthusiastic employees, set about this task with great energy and in close cooperation with the above-mentioned institutions.

The surviving portion of the factory mainly comprises the following units: the building for the production of nitroglycerine, consisting mainly of the nitration apparatus; the building, where the nitroglycerine was washed and the setting shed where gelatinizing took place; the bakery, where the compound was finally turned into dynamite substance. This building contains all the processing equipment for mixing and charging and illustrates the process of technical evolution from craft industry to mechanical production. A fully equipped laboratory has also been preserved. Adjoining the surviving buildings with their technical equipment is another building in which models and illustrations depict the history of explosives manufacturing in Norway. Finally, a separate building has been fitted out as an archive, containing valuable historical materials.

Dynamite played a tremendously important part in the development of Norwegian industry and communications. An efficient explosive was needed both for hydropower development and to construct roads and railways, and so it was found supremely justifiable to rank this facility as one of the most valuable from a Norwegian viewpoint. It should also be of great value in the international context, partly because the world's second nitroglycerine factory was built in Norway, and also because this, probably, is the sole surviving factory in the world to have been based on Nobel's own method.

The Bredalsholmen Drydock:
Preservation through Continuing Use

Danckert Monrad-Krohn

When work on the drydock at Bredalsholmen, outside Kristiansand, began in 1873, the underlying idea was nothing new. Plans for a drydock in Kristiansand had already existed in the 18th century, though apparently they lapsed into oblivion during the lean years after 1814. During the 1860s we read in the local press—*Christiandsands Stiftstidende*—that there is once again talk of a drydock. The transition from timber to iron and from sail to steam made the project more topical than ever. Over and over again, attention is drawn to the necessity and the possibilities of a drydock, and several attempts are made—unsuccessful—to raise the necessary capital. Bergen acquired its dry dock in 1858—the first in Norway—and drydocks were subsequently constructed in several coastal towns. When the Kristiansand project finally materialized, it was with the support of Bergen, and the facility was designed and construction work directed by the same engineer who had been responsible for the Bergen dock, N.H. Brunn, himself a citizen of Bergen. Work on Bredalsholmen began in 1873. The choice of location was no accident. The island occupied a central position in relation to shipping

The Bredalsholmen Drydock was built of stone on solid rock. Its original length was 97.5 meters, width 12.8 meters, and depth 5.5 meters, later enlarged to 137 × 24 × 6.25 meters.

lanes, a little way outside Kristiansand; the waters here were calm, and the topographical conditions were suitable. The dock, blasted out of a declivity in the terrain, was 97.5m long, 12.8m wide, and 5.5m deep. The bottom and sides were mostly solid rock. The dock was completed in 1876, and that same year, the Christiansand Dock Company admitted the first ship.

From the very outset, the dock was a popular excursion point for the local population, and on Sundays one of the small steamboats would usually carry trippers out the island to look at the ships being overhauled there.

In 1877, next to the dock, a small engineering works was established with the machinery for minor ship repairs. This comprised a lathe for shafts, a planing machine, a drilling machine, and a furnace for heat treatment and forge. Instead of operating this workshop directly, the company made it available to workshops using the dock. In addition, living quarters were built for the dock machine operator. Following a fire in 1882, a telephone link with Kristiansand—the first in the town—was established for safety reasons. The next fire, in 1887, broke out one day when the telephone was out of order!

In 1908 the dock was sold to Kristiansands Mekaniske Verksted, KMV, and over a year or so, starting in 1927, it was expanded to its present dimensions: 137m long, 24m wide, and 6.2m deep. This enlargement was partly financed by the municipality of Kristiansand to counteract the heavy unemployment at that time. After the Second World War, the old steam plant was replaced with an electric power supply from the mainland. New cranes were installed, the old workshop buildings were enlarged and amenities were erected in the form of a dining room, cloakroom facilities, showers, and toilets. At the same time, Bredalsholmen acquired a better link with the town, thanks to the construction of a road and bridge across the strait.

In many ways, time has stood still at

The whaleboats *Torlyn*, *Firern*, and *Thorarinn* docked at the Bredalsholmen Drydock in June 1934.

Bredalsholmen in recent years. The dock has been in operation all the time, but with little of the technical modernization that has taken place at similar facilities elsewhere, due to KMV's main commitments being to its shipyard in the city.

Bredalsholmen was sold in 1987 as part of a wider restructuring project, and its regular activities were discontinued. The municipality of Kristiansand has now taken over the property, with financial support from the Ministry for the Environment and the Norwegian Council for Cultural Affairs.

The dock in itself is an industrial landmark of great importance, associated with traditional economic activities of the town and region, viz, shipping and shipbuilding. Having changed relatively little through the ages, it is now unique by national standards. Moreover, the utilitarian potential of the dock in connection with the existing fleet of large veteran ships is immensely exciting. Bredalsholmen can be preserved as the industrial landmark it is, but also remain in active use as a repair yard for veteran ships. The shipyard is sufficiently equipped for this task, with cranes, pumps, compressors, gas supply, mechanical workshop, amenities, etc.

The Ministries of the Environment, and Cultural Affairs and Science have all pledged themselves to preserve a selection of old ships, including Norway's three large square-riggers, "the Training Ships." There are also many other ships, both timber and iron, that are too large to be drawn up on a slipway. Maintenance of the Bredalsholmen Drydock will be of the utmost importance for their preservation. Conversely, these vessels can make the dock economically viable.

There are several docks where ships of this kind can be repaired, but owing to the time-consuming working methods employed, e.g., clinkering instead of welding, repairs are long and often prohibitively expensive. For this reason, solutions have often had to be adopted that are contrary to principles of conservation.

Work is now in progress to establish a foundation that will take charge of running the dock. The aim is for operations to be self-financing. This is conditional on the foundation not incurring capital expenses and its expenditure being limited to running and maintaining the dock. The outgoing will have to be covered by dock and workshop rental payments from the users. Several of the veteran ships, e.g., the three training ships, have fixed budgets in which docking is included. Our preliminary inquiries lead us to believe that more commercially oriented concerns will be interested in using the dock and paying for it.

Clearly, then, we aim to do more than just preserve the Bredalsholmen Drydock as a landmark in industrial history. We also wish to contribute towards historically defensible ship conservation and, in this connection, to try and preserve necessary crafts and working methods now threatened with extinction. Clinkering is one example. Together with the coastal cultural center that has been opened on the neighboring island of Bragdøya and with the activities of the Agder coastal association at Grønningen lighthouse, Kristiansand will be placed well and truly on the "coastal cultural map."

The Industrial Labor Museum at Vemork:
A New Creation in an International Context

Bjørn Edvardsen

The Industrial Labor Museum, which opened its doors in August 1988, has resulted from cooperation between the trade union movement, the Norwegian hydropower authorities, and a private company. It illustrates what can be achieved by combined efforts for the pres-

The Industrial Labor Museum at Vemork opened in 1988.

ervation of historic monuments relating to working-class culture, power-intensive industry, and hydropower.

In 1975, for historical reasons, the Ministry of Industry ordered Norsk Hydro to preserve the old Vemork power station with its pipelines and appurtenant machinery. Construction work on the station and headrace tunnels together with the dam construction began in 1907, with 600 workers under the direction of eight engineers. In 1911 the power station delivered power for the first time from ten generator units, and with its 145,000 h.p., it was the biggest power station in the world—a position that it retained until 1916. Electricity production continued here until 1971.

The building was designed by the architect Olaf Nordhagen, and its architectural qualities, both inside and out, make it worth preserving. Today it is open as a museum, the aim of which is to illustrate a period of industrial development that gave people better living conditions for the most part and laid the foundations of social qualities from which we are still deriving great benefit. Large groups of the population grew up in industrial communities of this kind, and here we have acquired a new cultural institution that, by documenting operations and social conditions, will help to fill a gap in the national historiography.

Today about 75 percent of our museums of social history relate to the agrarian society. Most of them have been prevented by limited resources from taking new tasks upon themselves. The Industrial Labor Museum would not have been realized today if it had not been for the strong conviction on the part of the trade union movement that its members are also important carriers of a culture that must be recorded and preserved.

The Exhibition

Visitors today can see the first stage of the exhibitions planned for the museum, occupying about 400 square meters and dealing exclusively with the history of the industrial community of Rjukan between 1920 and 1945. This exhibition is based on ideas and themes devised by the museum itself. The permanent museum staff has collaborated with people representing various specialities and types of occupational experience.

The machine room illustrates "Hydropower—from Resource to Consumption," an instructive exhibition previously received by the museum from Norway's six largest electricity suppliers. The Information Service of the Norwegian Energy Authority has donated the exhibition "Light over the Land," a travel-

The machinery hall at the Industrial Labor Museum at Vemork showing turbines from 1911, at which time, the water power station was the largest in the world.

Photo: Fredrik Oftegro 1989.

ling exhibition constructed for the centenary celebrations in 1985.

The machine room—110 m long, 20 m wide, and 14 m high—is an exhibition in itself. It is being restored now to its condition in about 1930, and the reinstatement is nearing completion. Wall lamps and tiling for the floor and windowbacks had to be ordered specially abroad.

The technical installations have been altered and improved over the years. Ten twin turbines from A.G. Escher Wyss & Co., Zurich, and J.M. Voith, Heidenheim, installed in 1909, were fitted in 1918 with new blades that raised their capacity by 8 percent. The Voith turbines were replaced by Neyrpic, Grenoble, in 1954. All but one of them were reinstated as part of the restoration.

Until 1928, power to the Birkeland-Eydes arch furnaces at Rjukan was supplied by the three-phase A.C. generators from ASEA, Västerås, and from Brown Boveri & Co. of Baden, at 10–11,000 V and 50 periods. In 1927 it was decided to adopt the Haber-Bosch synthesis, which among other things meant large quantities of hydrogen. This process

Containers used in the production of heavy water.from the exhibition of industrial production in Rjukan 1920–40.

Photo: Curt Jacobsen.

required D.C., and so the generators were rebuilt in 1928, and a large hydrogen plant constructed overlooking the Vemork power station lake. This in turn meant a new by-product—heavy water, a topic dealt with in one of the other departments of the museum.

In future exhibitions we intend to show how the type of industry conducted at Rjukan affected the development of Norway. The Industrial Labor Museum is above all concerned with social history, while the history of technology is the special concern of the Norwegian Museum of Technology.

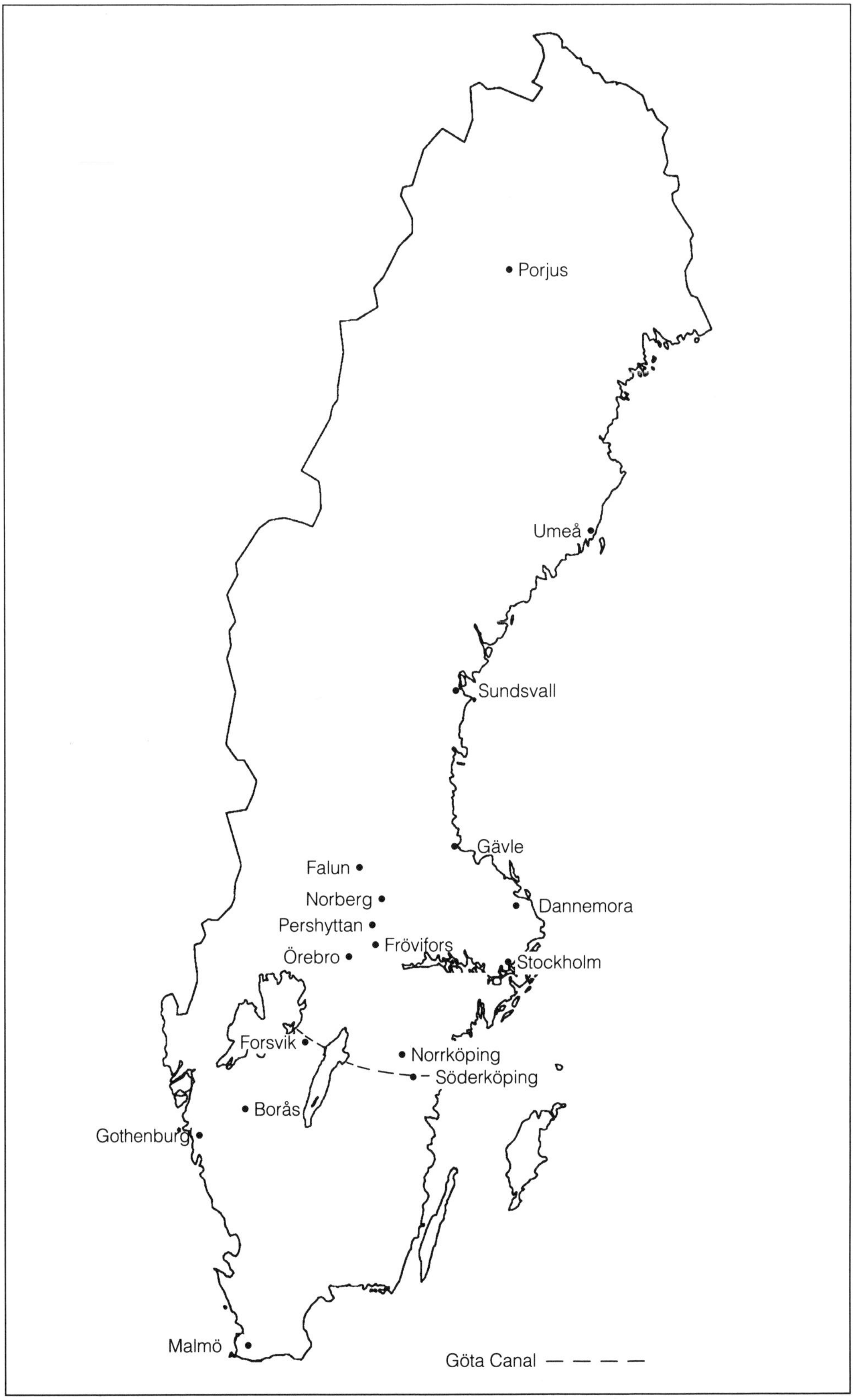

Porjus
Umeå
Sundsvall
Gävle
Falun
Norberg
Dannemora
Pershyttan
Frövifors
Örebro
Stockholm
Forsvik
Norrköping
Söderköping
Borås
Gothenburg
Malmö
Göta Canal

Sweden

Marie Nisser **Kersti Morger**

Sweden may be known for its vast woodlands, for its rivers and thousands of lakes, for its long and dark winter nights, and for the midnight sun making day out of night. This long country measures 1,574 kilometers from the south to the north, and the total land area is 449,964 square km. More than half of that area or 56 percent is covered by forests, and around 30 percent by mountains, heaths, bogs, etc. Only about 14 percent is cultivated land. Most of the population of some 8.6 million inhabitants are to be found in the southern part of the country.

The Rise of the Iron Industry in Bergslagen

Natural resources have been of the utmost importance for the growth of Swedish industry. Iron was produced in prehistoric Sweden. During the Iron Age, widespread production of iron developed based on the extraction of lake and bog iron ore. Primitive iron-making was mainly confined to areas where lake and bog ore are present.

More than 4,000 remains of early iron production have been included in the central Ancient Monuments Register. In many of those areas, such as the northern part of Dalarna and in Småland, iron-making using lake and bog iron has survived over the centuries. In Småland this long-lived tradition did not end until the 1930s. Some of the well-preserved industrial sites and buildings remind us of the iron production of those bygone days, among them the charcoal blast furnace at Åminne and Huseby ironwork's estate.

Large numbers of valuable and fascinating minerals in the soil and rock were to become extremely valuable for Sweden. It is not known for sure when rock mining started. Archaeologists have wanted to predate the discovery of the copper mine at Falun to the eighth century, A.D., or perhaps even earlier. And the metallurgical excavations of Lapphyttan indicate that the blast furnace process was practiced in Sweden by the 12th century. From that period, the Bergslagen area, with its expanding mining and iron industry, was of increasing importance for the

Swedish economy (see the article on Bergslagen in this volume). In fact, Stockholm, the capital of Sweden and the port between Lake Mälaren and the Baltic Sea, derived much of its prosperity and rising importance during the Middle Ages from being the transit-harbor for Swedish iron.

The Bergslagen region witnessed the quick expansion of the iron industry, the short-lived but early important period of the Sala silver mine during the 16th century, and the glorious period of the Falun copper mine in the 17th century.

The dramatic growth of Swedish iron production and exports in the same century is unmatched in early preindustrial expansion. Up to the mid-18th century, Sweden was undoubtedly the biggest iron exporter in Europe, with a large share of the international iron trade. The Bergslagen area was also the birthplace of the modern iron and steel industry in the late 19th century.

The countryside where iron-making developed was idyllic. The Nordic light gave it a mysterious appearance with distant blue mountains that hid the ore, deep and dark forests that provided fuel for the blast furnaces and forges, glittering waterfalls that gave the power to run the bellows. The small industrial units were in remote and sparsely populated areas. The colorful owners who often ran them developed their own cultural and economic milieu. To a large extent, such factors often led to the cultivation of technology and natural science in Sweden.

Expansion of the Forest Industry

The forest industry expanded early. In the 16th century, a considerable amount of timber for masts, board, and firewoods was already being exported from the coastal provinces. During the reign of Gustavus Vasa, the burning of tar and shipbuilding experienced a notable upswing. The export figures for tar were only exceeded by iron, copper, and silver.

In the 17th century, water power could be adapted to run the sawmills. Production grew rapidly, as did the exports of wood products. However, much of the timber was also needed for domestic purposes. Vernacular buildings and most of the houses in the small towns were built of wood. The competition for woodlands hardened between the ironworks and the sawmills.

Nevertheless, the sawmill industry grew considerably in the 18th century and gained much in quality by the innovation of the fine-blade saw. From this century onwards, the state issued rules and regulations to protect the woods from deforestation.

Sweden was a major participant in European wars between the mid-16th and mid-17th centuries and as a result, experienced a considerable territorial expansion. Warfare required much organization. The naval dockyards at Stockholm and the naval city of Karlskrona were big and important workplaces during this period.

The Paper Industry

The paper industry had a late start in Sweden compared to Spain, Italy, and France. The first papermills in our country were not founded until the late 16th century. The initial stage was slow but growth accelerated in the 18th and early 19th centuries. There was a notable concentration of papermills in southeast Sweden, especially in the provinces of Småland and Östergötland. Fifty percent of the roughly 100 hand papermills that have been in operation in Sweden were located in Småland. Handpaper is still being fabricated there at Lessebo bruk. At Ösjöfors papermill the old workhouse with its antiquated installations and the manor house of the papermaster and his family remain almost intact since the mill was closed in 1923. This is a fascinating preindustrial site and is an excellent illustration of hand papermaking before the age of mechanization.

Proto-Industrial Regions

According to the traditional view, Swedish industrialization occurred late compared to certain other industrializing nations. Economic historians have, however, recently questioned this model of Swedish industrialization. The agrarian revolution in the late 18th and early 19th centuries and its impact on society have been a matter of discussion. During that period there was an enormous amount of land reclamation. The systems of rotation were improved. Better drainage was introduced; new tools and methods came into use. Agriculture played a crucial role in the establishment of a growing home sector. Capital was most probably transferred through the credit system from agriculture to industry, but the extent of this contribution is still unknown. Attention has been drawn to other important factors contributing to industrialization, such as rural crafts and artisanal production. The importance of a growing home market has also been emphasized in several studies.

Dalarna was the most important province for proto-industry. Every type of handicraft was practiced within the province. Wood and metal crafts were particularly widespread. Many small workshops for woodcarving or making scythes, for example, can still be found in the villages in this region.

In many parts of Småland, the rural industry was of a high standard—wire-making, the bayonet industry, coopers, carpenters, clockmakers, and many others. It was from this craft industry that an extensive factory industry developed in the 19th century. Småland became a center of wood industries, such as sawmills, furniture making, and pulp and paper mills. In the county of Kronoberg, the production of linen was particularly well developed. It was also in this region that the glass industry became of growing importance in the 19th century. Glass making has preserved its handicraft structure and is still of great importance in this district, being one of the most successful sources of industrial tourism, especially in the summer.

There were, however, other significant proto-industrial areas, for example, southern Västergötland—Sjuhäradsbygden—with its textiles, wood, and metal products. Already in the Middle Ages the region around Borås had become an important center of domestic spinning and weaving, predominantly based on linen. During the 18th and 19th centuries, handloom weaving became a more and more specialized and organized domestic craft in this area. Middlemen-entrepreneurs were used as organizers. Just before the takeoff of industrialization, thousands of domestic weavers were handled by some 40 middlemen-entrepreneurs. The trade in linen and woolen cloths that laid the foundation of economic growth and wealth can be traced in a number of manor houses from the 19th century.

The Industrialization of Sweden

From the 1750s to the 1850s, the population of Sweden grew from 1.8 million to 3.5 million inhabitants. In 1850 nearly three-fourths of the Swedes depended on agriculture for their livelihood, and less than 13 percent lived in the small and still rather unhealthy cities. The whole sector of mining and manufacturing, including all sorts of artisans and craftsmen, did not amount to more than one-seventh of the population. Industrial workers may have numbered 8 percent of the active males.

The transformation of the industrial scene in the late 19th century was dramatic. New methods were introduced in the iron industry. The modernization of the blast-furnace process was based on the adoption of foreign innovations. The blast furnace at Långshyttan in Dalarna built in 1859 by Håkan Steffansson started the modern era of pig iron production. Production capacity was increased, and fuel consumption decreased. The charcoal process died hard, however. In 1950 there were still some 30 charcoal blast furnaces producing pig iron, and the last charcoal blast furnace was only closed down in the 1960s.

The 1859 blast-furnace buildings at Långshyttan are still preserved next to an older, timber-clad blast furnace rebuilt in 1759. Several "modern" blast furnaces from the late 19th and early 20th centuries are among the most outstanding monuments of our industrial heritage.

Iggesund blast furnace rebuilt in the late 19th century and closed down the the 20th century. It is scheduled as an industrial monument. *Photo:* Marie Nisser.

In 1845 Gustaf Ekman built a welding furnace with a gas generator fueled by charcoal. His method was the so-called Lancashire process. It replaced the German forging that was a time-consuming process based on handicraft skill for making bar iron. New capacity in production was gained, and the quality of the Lancashire bar iron was excellent. The great expansion of the Lancashire process started in the 1860s. Lancashire iron was being produced until the 1960s, and the last Lancashire forge and rolling mill to close down was Ramnäs in the county of Västmanland in 1968. It has been preserved as an industrial monument. There are, however, a few more Lancashire forges conserved *in situ*, and the one at Karlholms bruk in Uppland and the Wij former rolling mill in Gästrikland are two other good examples.

In the same period, Göran Fredrik Göransson bought one-fifth of Henry Bessemer's patent for Sweden and started the first experiments at the Edsken blast furnace in November 1857. After several trials he managed to produce the first charge of excellent steel on 18 July, 1858. He had altered the converter, mainly by enlarging the apertures to get higher air pressure during the process. He also got the best results when he used iron ore with some manganese content. In the autumn of the same year, he visited England and demonstrated ingots of high quality from Edsken. Henry Bessemer then started to produce Bessemer steel in Sheffield.

Göransson became the manager of the new steelworks at Sandviken. Some ironworks were quick to start producing steel according to the new invention. But the great expansion of Bessemer works reached its peak in the 1870s. The Bessemer converters at Hagfors are conserved *in situ*, perhaps the only ones of their kind.

Of even greater importance to the Swedish steelmakers was the introduction of the Siemens-Martin open-hearth process patented in 1867. A similar innovation at the same time was made by Johan Fredrik Lundin at Munkfors. The steelworks at Munkfors have moved to modern premises, but the former open-hearth works with its installations from the late 19th century has been preserved as a museum.

The restructuring of the iron industry in the late 19th century was a hard-striking process. The smaller ironworks were soon ousted by the larger units with modern equipment. The older ironworks with their outdated manufacturing methods were often in remote places, far from modern transport facilities. Hundreds of small blast furnaces and forges were forced to close. By 1913 only 177 out of 213 blast furnaces that existed in 1870 survived and 140 ironworks out of 381.

Perhaps even more spectacular than the introduction of steel-making was the expansion of the sawmills along the coast of Norrland. The demand for timber and boards had risen dramatically during the industrialization of Western Europe. The inland forests of Norrland lay open to exploitation as soon as the large rivers had been cleaned to make log-driving possible. Steam power could be used to increase production in the sawmills. The first steam-powered sawmill was built in 1847, and roughly 20 years later, their number had increased to 100. During the same period, the Swedish export of boards and deals multiplied five times. Around the 1870s it was the leading export industry. However, the market situation changed rapidly; demand declined; and good timber for boards became scarce at the turn of the century. The expansion of the pulp and paper industry along the coast of Norrland from the 1910s onwards helped many of the new industrial communities survive.

The sawmills of that period were bigger than the ironworks, but the general structure of each was much the same. The technology of the installations was comparatively simple. Work in the forest and floating timber were largely a seasonal sideline for many farmers. Little remains of the sawmills and the workers' housing from the period. The sawmills were built of wood and disappeared soon after they closed. The housing had been rapidly and poorly built. No one seemed to wish to preserve these buildings that reminded people of the poor working-class conditions during the early period of industrialization.

Industrial production in the textile industry started in 1834 in the southwest—Sjuhäradsbygden (mentioned earlier) with its widespread domestic textile production. In that year Sven Erikson built the cotton-weaving mill at Rydboholm. It was the first factory for power weaving in Sweden. Twenty years later it proved necessary to build a cotton-spinning mill to supply Rydboholm with cotton yarn. That mill was built at Rydal nearby. The mill had 350 employees in 1953, but since then the number has steadily diminished. In 1984 the original company went into liquidation. In the following year, a museum project was launched to make Rydal a textile heritage center.

Norrköping in Östergötland took the lead in the 1850s with the rapid growth of new cotton and woollen mills along the Motala river in the city center (see the article on those mills). In the 1870s the city of Borås in southwest Sweden started to expand when two weaving mills were built in the city center, and they were followed by others in the 1890s. Most of the hundreds of textile mills that settled in Borås were cotton mills, knitting mills, and ready-made clothing factories.

Railway building started slowly but experienced decisive growth in the 1890s. A number of stations from the early days of railway expansion survive to illustrate the history of Swedish transport.

The great wave of European economic expansion in the early 1870s led to astonishing industrial growth in Sweden, with considerable effects on the forest industry with the breakthrough in the pulp industry, but also in iron and steel-making and in the increasing importance of engineering.

Industrial Sweden

The transformation that first made a really industrial country out of agrarian Sweden occurred during the 20 years from the middle of the 1890s to the First World War. This period was the beginning of a new era in Swedish industrial history. Hitherto, Sweden had been able to profit from domestic natural resources. To a great extent, its industry had developed and expanded as a result of foreign demand. A new factor was now added—a degree of creativeness leading to new inventions that opened another field of progress. Lars Magnus Ericsson designed the first desk telephone only a couple of years after Bell's invention. Jonas Wenström developed the three-phase system for the transmission of electric currents, and ASEA was founded in 1883. Sven Wingquist invented the self-regulated ball bearing, and SKF started production in Gothenburg in 1907. C.E. Johansson invented the famous "Jo-block" precision guages. Gustav de Laval invented the milk separator 1877, and Alfa-Laval started production in 1883. Gustav Dalén was later to receive the Nobel prize for his world-renowned automatic navigation beacon, which led to the foundation of AGA. These "elite" companies represented a small part of Swedish manufacturing before 1914. The new initiatives helped, however, to strengthen Sweden's reputation abroad as a nation of technicians and gave Sweden a stronger position among the industrializing countries. The firms developed into multinational companies at a remarkably early stage.

Many of the factory buildings of those pioneering firms are still to be seen. They represent a fine tradition of industrial architecture of the early 20th century. Such examples are the ASEA workshops in the city of Västerås and SKF's factory in Gothenburg.

The development of the three-phase system enabled power to be transmitted over long distances. Sweden was swift to exploit its waterfalls. Hydroelectric power stations were built all over the country, and the government became increasingly involved in the development of national hydropower. The importance of this branch and some of the notable hydroelectric powerstations built in the early 20th century are mentioned in the article on Porjus.

The iron ore deposits at Kiruna were known in the 19th century to be among the greatest in the world. Mining operations on a larger scale could, however, not start until the transport problem had been solved. A mining company was formed in 1890 named after the two mountains Luossavaara-Kirunavaara, abbreviated to LKAB. In 1898 Parliament voted to build a railway from Luleå on the northeastern coast of Sweden passing Kiruna to Ofoten on the western coast of Norway. This opened the possibility of extracting the rich iron ore deposits. The population in the mining village grew rapidly from just over 200 in 1900 to about 7,000 in 1909. A shantytown grew up, built by the miners. The company saw the need for Kiruna to become a planned community. Two Stockholm architects were commissioned for planning and building. The community was planned for 500 families living in different types of houses designed for workers, and each family had its own garden. This frontier city is an important contribution to the industrial heritage of Sweden.

Along the coast of Norrland, the industrial landscape was transformed. Many of the sawmills were replaced by pulp mills. They were larger than their predecessors and employed more people. They also generated larger and more elaborate communities with better quality housing for the skilled workers than the sawmill communities had.

The continuation of the prewar expansion was not without its problems and disappointments. The depression in the early twenties was especially severe, and there was a good deal of unemployment—but in comparison to the general European development, the net result was unusually positive. Sweden was among the few nations where the 1920s as well as the 1930s produced considerable growth. It was also a period of technical and structural changes. This transformation was to be of fundamental importance for later expansion. Sweden was extremely fortunate in not being involved in the two world wars. In the interwar period, the market situation was skillfully exploited for Swedish products. Undamaged factories could increase their production for an expanding market without any delay.

The last century has seen an enormous decrease in agricultural employment and a strong increase among industrial workers. Agriculture was the greatest employer in Sweden up to the late 1930s, when it was surpassed by industry, which in turn was supplanted by services during the 1960s. Seven hundred thousand persons were still employed in agriculture after the Second World War, but by the 1970s, their number had decreased to 20,000.

Already in the 1950s, the textile, clothing, and shoe industries experienced a regression, and during the following decade, their decline continued. The textile mills in Norrköping, Borås, and elsewhere had to close, as did most of the shoe factories at Örebro. The town planners were confronted with the problem of big industrial areas in disuse in the city centers. Land reclamation and replanning seemed to be the only answer in the 1960s. It was not until the late 1970s that adaptive reuse started to be discussed.

Other industries, however, witnessed a rapid development in those years, especially the engineering industries, such as Electrolux, SAAB, ASEA, Atlas Copco, and Alfa-Laval.

During the 1970s the structural problems of Swedish industry became more and more evident. Shipbuilding and mining were particularly hard struck. Sweden had nearly 10 percent of the world shipping tonnage. But like other shipbuilding nations in Western Europe, our country could not withstand competition from nations capable of buildings ships at lower costs. Most of the large shipyards in Gothenburg, Malmö, and Landskrona were forced to closed down with severe affects on employment and economic stability in those communities. In Gothenburg the shipyards have been recorded, and preservation projects have been launched.

Today, the mining industry is extinct in the Bergslagen region. The Dannemora mine closed in 1992 after having been in operation for more than 500 years, and so has the famous copper mine at Falun. The mining history of Sweden is still waiting to be written, and a plan for future conservation must be drawn up.

Only 12 steel-melting plants are still in operation. In addition, the rolling of steel is undertaken at nine places. The restructuring and continuous modernization of the steel industry have been going on throughout this century. Special steel, i.e., alloy and high-carbon steels, comprise 45 percent of Sweden's total crude steel production, a higher proportion than in other major steel-producing countries. The total production in terms of tonnage is relatively small. Nonetheless, the manufacturers often assume a prominent position in both the domestic and international markets.

The paper and pulp industry has also undergone a far-reaching transition during the last decades. More than 50 percent of the pulp and paper mills that operated in the 1950s have closed down. The modernization and expansion of the working pulp and paper plants have, on the other hand, been extensive. Companies have fused, and a few large mills now dominate the market.

Domestic natural resources do not play the role they did during the earlier periods of industrialization. Our future chances will depend on our inventiveness, our technical skill, and experience. This change in our position makes the need for industrial restructuring far more important.

The Conservation of Sweden's Industrial Heritage.

Some provision for the care and documentation of the industrial heritage has existed since the turn of the century. This has mainly been confined to one branch of the economy—the iron and steel industry. In this connection it may be of interest to note some of the most characteristic features of the beginnings of the conservation.

It is no coincidence that the former industrial sites of the iron and steel industry were chosen for conservation. Hundreds of charcoal blast furnaces and forges with their old-fashioned methods and small production capacity were still at work in the late 19th century. Bergslagen, where the iron industry over the centuries has left its impression on both landscape and building, was by 1900 being completely transformed. The structural changes of furnaces and forges in Bergslagen were, however, only part of a process of industrialization whose pace had accelerated at

Pershyttan blast furnace, one of the well-preserved charcoal blast furnaces with adjoining peasant miners village. Production stopped in the 1950s. Photo from the turn of the century shows workers outside the furnace.

Source: the Swedish National Museum of Science and Technology.

the end of the 19th century, and whose affect on Swedish society at large became more and more marked.

The old cultural regions were losing their former role in the industrial life of Sweden. In poetry, art, and architecture, however, there was a return to these regions, rich in memories and fabled environments. The wave of national romanticism that swept the country in the late 19th century was often inspired by the world of ironmasters and smiths.

Managing directors and local historical societies succeeded in guaranteeing the future maintenance of several of the charcoal blast furnaces and forges that stopped production around the turn of the century. The furnaces and forges were often owned by large companies that accepted the responsibility of continued care and maintenance. Companies, such as Stora, the former Uddeholms AB, and Iggesunds AB, have conserved many industrial sites connected with the production of iron and steel. There are numerous reasons for this. Many of the managing directors of the old ironworks had grown up in these surroundings. For them, historical continuity was obvious. Another factor was the publicity value attached to these industrial monuments. Stora Kopparberg took pleasure in showing its customers and other visitors the copper mine at Falun, which for nearly 1,000 years had been the center of the company's operations. The blast furnaces and forges, virtually unchanged in form, appearance, and equipment from the 1620s to the 1820s, provided a fitting illustration of the Swedish iron industry and its proud traditions.

The preindustrial works that have been preserved and maintained were until the 1970s primarily those dating from the early period of the iron industry. Financial considerations play an important part here. Only a few buildings dating from the later phases of the industrialization of Sweden are preserved as monuments. And yet there are a few such sites of great importance because they seem to be unique in the context of the World Industrial Heritage. On a small island in Lake Åmänningen at Ängelsberg in the province of Västmanland, a unique petroleum refinery has been preserved. It seems to be the oldest conserved plant of its kind in the world and is of the same type as the first petroleum refineries built in the U.S.A. in the 1860s. The well-preserved open hearth at Munkfors in Värmland dating from the late 19th century is another example. Still another worth mentioning is the Bessemer plant preserved *in situ* at Hagfors in the same province. Some impressive late charcoal blast furnaces that were still operating in the early 20th century are also being preserved. The blast furnaces at Moviken and Iggesund in the county of Gävleborg and Avesta blast furnace in Dalarna should also be mentioned.

Museums for the History of Technology and Industry

The interest in industrial heritage conservation and documentation also spread to other industries in the first decades of the 20th century. Among them were the paper and pulp industry where steps were taken to conserve some of the older plants.

The Jubilee Exhibition in Gothenburg in 1923 was of great importance in this context; it was characterized as a mustering of "Swedish life and work." There was plenty of room for objects connected with industrial history. In 1924, the Museum of Science and Technology in Stockholm was founded. Its collection was built up under the direction of Torsten Althin. His work commands respect, resulting as it has in a wide collection of objects, impressive archives, and an extensive library. The museum has always emphasized the documentation of engineering history. Other museums with collections illustrating the history of science and technology came later. One of the most important was the Industrial Museum in Gothenburg.

Several company museums were also opened during the first half of the twentieth century. Stora Gruvstugan (the Great Mine House) at Falun mine, built between 1771 and 1785, was restored by Stora Kopparberg and

The governor of Gävleborg county, Sven Lübeck, applying a commemorative tablet in 1932 to the engine house at the Dannemora mines in Uppland. The building was constructed in the 1720s by Mårten Triewald to house the steam-powered hoisting machinery fashioned according to designs by Thomas Newcomen.

Photo: the Swedish National Museum of Science and Technology.

opened as a museum as early as 1922. Surahammar opened its ironworks museum in the authentic surroundings of the old forge in the 1950s, to name just two of the major company museums.

In Ludvika in the center of Bergslagen, a mining open-air museum was opened in the late 1920s. It was probably the first industrial open-air museum in the world. During the following years, it was extended by adding new objects. From the start the museum has been run by the local historical society in Ludvika.

The survey "Workers' Memories" that Nordiska Museet (the Nordic Museum) carried out in the 1940s with the aid of a broad-based series of interviews and questionnaires was an excellent complement to the technological approach that had dominated the museums just mentioned. It resulted in a great collection of memories that today is considered a valuable source for the history of the working environment in Sweden. Some of the interviews were published: "In all its changing multiplicity, a credible and gripping retrospect of the rise of modern industrialized society," as a reviewer described them at the time of publication.

A sort of ecomuseum before the term was invented was Husbyringen, which sought to portray an entire region with its nature, animal life, cultural heritage, and not least its industrial heritage. The 1980s saw the creation of several other ecomuseums, among them the Ekomuseum Bergslagen.

Today there are more than 650 industrial sites, museums, exhibitions, and collections that illustrate the working heritage of our country and in particular document the growing importance of the industrial heritage as part of our cultural heritage.

Industrial Heritage Studies in the Last Two Decades

The early efforts mentioned above were isolated incidents where private initiative was decisive. In the 1970s, however, the situation changed. Industrial sites and their environment became interesting, not only to a few historians and antiquarians, but also to study groups and a broader public.

The voluntary efforts of private persons and associations in many parts of the country are a vital part of the work now in progress. Then again, many companies have put money into documentation, revival, and reuse. Important and substantial contributions have also been made in this connection by various organizations—the Swedish Ironmasters' Association, the Forest Industries Association, and the State Power Board.

Many factors contributed to the change in attitude. The structural transformation of Swedish industry struck hard in that decade. Industries closed down, factories were emptied, and industrial villages depopulated. It suddenly became important to many persons to preserve the links with the past. This in its turn created an attitude favorable to protecting the remains of the industrial heritage.

The environment of everyday life—work places and housing—started to be regarded as important for its identification with the past. Suddenly this heritage became part of the cultural heritage. A decision taken by Parliament in the mid-1970s stated that the cultural heritage should be protected in this broader perspective.

The Swedish Worker's Educational Association set up study circles all over the country, and they have been extremely active in the last two decades. The idea was to encourage working people, either individually or in study groups, to write or compile their own history. Exhibitions were arranged; reports and books were published. Amateurs wrote plays about the workers' conditions and their political struggle and the formation of the labor movement. At this time the discussions started about the need for a museum of work, and that museum was inaugurated in a former textile mill at Norrköping in December 1991.

As many as 10,000 study circles were active from the mid-1970s to the mid-1980s. Thousands of persons studied their industrial heritage and their own past. More than 400 publications were printed, and plays were performed at around 100 different places.

Industrial Heritage Conservation in the Last Two Decades

In the 1970s national planning was carried out to improve the use of resources, land, and water; preservation interests were given an important role in this. So far, thorough surveys have been carried out in each county, and attempts have been made to arrive to evaluate buildings and sites of cultural historical value. Local authorities throughout the country have sought to protect monuments and areas of national interest. Among these are, of course, many industrial establishments. It is, however, mainly the early structures that have long been held to be irreplaceable. Continued planning demands that the basis of the survey be extended to cover 20th-century environments, so that we may through preserved objects have a more complete picture of the industrial development of the country.

As has previously been mentioned, the tight financial limits within which building conservation has to work have long been an obstacle to broadening the present aims. An exception—and an important one—was the period from the late 1960s to the early 1970s when the Swedish Labor Market Board made a substantial contribution to the repair and care of many historically important buildings and environments. Many construction workers who were unemployed at that time thus helped restore old buildings. Several industrial monuments were restored. The Falun copper mine was adapted for tourism, as were Engelsberg's bruk in the province of Västmanland, the Lancashire forge at Karlholm in Uppland, and the Moviken blast furnace in Hälsingland, to mention only a few. But the situation changed in the mid-1970s.

Trusts and nonprofit-making institutions could more easily raise money or assistance from the Labor Market Board than private owners. Against this background, Föreningen Värmländska Industriminnesmärken (the Society for the Protection of Industrial Monuments in Värmland) was founded in 1972. Its aim was to foster research by preserving industrial monuments and making them accessible to the public. Private owners, generally the larger companies in the county, have transferred many of their industrial buildings and sites to the society. Among such historical industrial plants can be mentioned several blast furnaces, the Munkfors open-hearth furnace, the Bessemer plant at Hagfors mentioned earlier, and some water-powered grain mills.

Adapted reuse of empty industrial buildings that have lost their original function started to be regarded as a possible solution at the beginning of the 1980s. Many factory buildings from the early Industrial Revolution are of notable architectural quality, in addition to being painstakingly and solidly built. Their dimensions and general layout tell us of important stages in the development of industry. One example is the industrial landscape along the Motala river in Norrköping where several of the former textile mills have been adapted for other purposes. One of the latest examples of successful reconversion is the Museum of Labor at Norrköping. One of the most interesting mills in Borås is now being altered to house the Textile History Museum. The museum will be inaugurated in the new premises in 1992. This cotton-spinning mill of the former Åkerlund company was built by the English architect Philip Sidney Stott (1858–1937). Many of the other fine textile mills in southern Sweden and some of the steel mills at Eskilstuna have been converted for adaptive reuse.

Planning and Legislation

When describing the aid available for the protection, care, and preservation of landmarks of industrial significance, it is interesting to study the legislation that has been passed in Sweden since 1987. The prevailing objective since the 1970s has been to integrate care of the architectural heritage with the fabric of urban planning. In shaping the fiscal environment and in connection with changes of land use, sociohistorical values are also taken into account. This policy is expressly declared in the Resource Management Act that came into force on 1 July, 1987.

Another act worth mentioning is the Planning and Building Act (PBL), that came into force simultaneously with NRL. PBL makes the municipalities expressly responsible for harmonizing land use and urban planning with the interests of historical preservation. It lays down safeguards "for all buildings and constructions that are particularly valuable from a historical, cultural, environmental, or artistic point of view or that are part of an area of buildings of such value."

Statutory provisions relating to management of the cultural heritage have been codified in a new act that provides a much more coherent overview than used to be the case. Until recently separate acts governed each of the four main categories of the cultural heritage in Sweden. In 1988 these were merged into a single act concerning ancient monuments and finds (KML) that took effect on 1 January, 1989. Of particular relevance to industrial landmarks are the statements made in a government bill introduced before KML. Among other things that bill states that the term "cultural heritage management" is not to be confined to the specifically historic parts of the cultural heritage: "The nucleus of this work should be an effort to utilize historic values in integrated environments in the man-made landscape, in urban settlements, and in industrial development." These points testify to the lofty aims that have been defined for cultural heritage management. Industrial landmarks constitute no small part of that heritage and their documentation and preservation can therefore be given greater attention.

From this account it might seem as if industrial heritage conservation had developed without any problems, but unfortunately there is another side to the picture. There are often financial problems to be contended with, and the difficulties can seem insuperable, in view of the formidable financial inputs needed. However, an account will now be given of measures and projects, showing the diversity of initiatives and efforts deployed in this field.

Inventories, Preservation, Documentation, and Revival of the Industrial Heritage—a General Account

Various inventories with a bearing on industrial heritage management are in progress at the Central Board of National Antiquities. Since the beginning of the 1980s, the Archaeological Documentation Unit of the Ancient Monuments Department has undertaken systematic inventories of industrial-preindustrial sites. Even before that, in connection with the regular inventory of ancient monuments, records were kept of the remains of iron and copper furnaces. From 1980 industrial remains in the broad sense have come to be registered as part of the regular inventory of ancient monuments for the economic map of Sweden. This has led to thousands of mine shafts being registered, plotted on the economic map, and described. Inventories have also been compiled of many mining areas.

The extensive archaeological investigations of the medieval furnace site at Lapphyttan in Norberg have made it possible to distinguish a medieval blast furnace horizon in a large corpus of furnace remains comprising more than 700 sites. There may be roughly 20 medieval furnace sites extant in the whole of Sweden. This is a remarkably high figure. Archaeological investigations of these sites would be feasible, but the cost is prohibitive. Most of the medieval furnace sites have been reused since the Middle Ages, in which case they usually present furnace ruins from the 18th or 19th centuries.

Studies of the material from the medieval furnace site at Lapphyttan have been going on since the mid-1980s. Work is in progress on reconstructing the entire furnace site. In connection with this reconstruction of the furnace, several interpretations of the archaeological material will be put to the test.

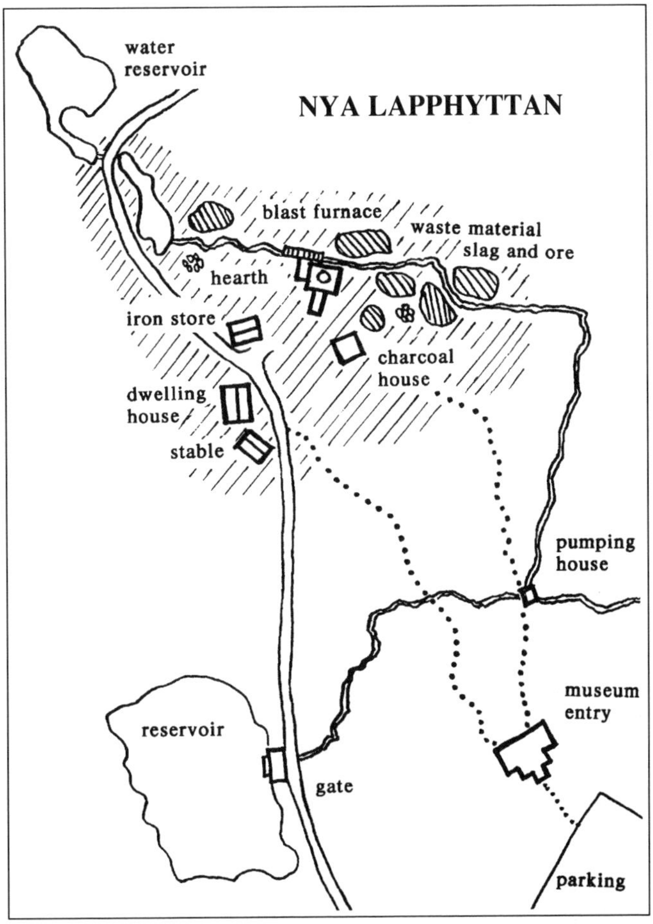

Nya Lapphyttan, Norberg: the reconstructed situation plan of the old Lapphyttan.

Three major inventories are in progress in the Buildings Department of the Central Board of National Antiquities. One of them comprises preserved mines and appurtenant facilities and buildings, a second gasworks, and a third power stations.

Extensive inventories of industrial buildings and sites have been compiled at county and municipal museums, providing input documentation for cultural heritage management program, preservation program, and conservation measures. In many counties these inventories are quite comprehensive. Categorized inventories are also common. One example illustrates this. In Västernorrland, a heavily forested county in the far north of Sweden, an inventory has been compiled of all forest-based industry. Historic industrial sites from the first ventures of the 16th century down to the time of the inventory have been charted and documented. The inventory comprises water-powered saws, ironworks, glassworks, steam-powered saws, electrically powered saws, pulp mills, and paper mills.

Every county in Sweden can present a large number of industrial environments, and many of them, following documentation, have been allotted a preservation program and made the subject of conservation measures. The central board has assisted with planning work at some of them, but in many cases, the conservation programs are drawn up by the county museums.

Within SAMDOK, which stands for contemporary documentation by museums and has an ethnological slant, a number of interesting projects are under way. Various pools are operating, with various specialities. Teko Museum in Borås, for example, has documented Swedish artificial silk. Axe production at Hults Bruk, the sawmill at Edane, the Ala export sawmill, the shoe industry, Vittinge brickworks, and other interesting industrial concerns have also been documented.

Industrial organizations are also engaged in extensive inventory, documentation, and research activities through the historical committees they have set up. Those committees include the Metallurgical History Group of the Swedish Steel Producers Association, which has started a national inventory of facilities in industry. This group includes researchers from various disciplines, metallurgists, and other industrial representatives, who are undertaking wide-ranging research on an interdisciplinary basis. For example, upwards of 40 metallurgy and geotechnology graduates, together with other representatives of the industry, are now documenting technical developments from 1850 to the present. The Swedish Pulp and Paper Association celebrated its centenary in 1990. For several years now the association's Historical Committee has been conducting a national inventory of abandoned mills and of pulp mills and paper mills that are still in operation. The first results of this documentation will be presented in four publications describing how the industry has developed in four regions of Sweden. The State Power Board has set up a Cultural Heritage Committee that is promoting the documentation of power industry facilities and also the compilation of a conservation plan for installations of historical interest that should be preserved for the future. Several publications on power stations have appeared already, and others are planned. The regional units of the State Power Board have now taken charge of documentation and conservation activities, with the Cultural Heritage Committee retaining general responsibility.

Finance

The refurbishment and care of industrial landmarks are often a financial burden. Both individual industrial enterprises and sectional organizations, as well as the Swedish Trade Union Confederation (LO), have recently invested large sums to preserve industrial landmarks and are continuing to do so. Industrial enterprises have a long-standing responsibility for many industrial environments of national and international interest. Over the years they have cared for these facilities and, all in all, are still doing much in this respect. Both the Swedish Pulp and Paper Association and the Swedish Paper Workers' Union have made generous donations to preserve the old paper mill at Frövifors. LO has made generous grants to the Museum of Labor in Norrköping, to quote just two examples.

State and county administrations, municipalities, and private companies can obtain state grants for the restoration, on a temporary public employment basis, of places of historical interest. Grants of this kind, however, can only be awarded when justified by the employment situation, e.g., if there is heavy unemployment among construction workers. Although it has been more difficult to organize temporary public employment projects and to finance big restoration undertakings, important measures have been taken in recent years. Pride of place should perhaps go to the expansion by the State Power Board of the Älvkarleby power station, and the board's decision in this connection to keep the old power station, built in the 1910s, intact and construct a completely new installation at a different site.

As part of a vigorous job creation scheme, the allocation for conservation of the cultural heritage was given a one-off reinforcement 1991 for investment projects relating to important cultural landmarks and archaeological sites in need of refurbishing. Roughly half this allocation referred to industrial environments.

But state grants are also payable for the care of urban settlements of historical interest. These loans and grants, of course, are a great help, but they are not generally sufficient to cover the full cost of restoring and caring for industrial environments, most of which are both large and expensive. A special government commission has been studying the administration of certain cultural environments, one of its tasks being to look for possibilities of preserving castle, manor house, and factory environments.

The final report, presented in July 1991, points to the crucial differences between the two categories investigated, as regards both applications and conditions of ownership, etc. The differences are conspicuous. The investigator found that the conservation of industrial buildings and sites is perhaps the most burning question of management in the conservation of the architectural heritage, as regards both the selection of items for preservation and the ways of safeguarding them, and also as regards maintenance for their refurbishment and subsequent care as landmarks of industrial history. The investigator firmly believed that the problems of industrial heritage conservation are as great as to require a separate investigation. He went on to say that ambitious, high-quality conservation of industrial landmarks, coupled with an active search for future uses, demands not only resources for operation and maintenance, but also resources for greatly improving accessibility to the general public and for bringing the landmarks to life.

Another of the distinctive characteristics of the conservation of industrial buildings and sites is that the number of items is increasing constantly and rapidly—in principle, at the same rate as Swedish industry is being restructured and transformed. This often involves heavy-process industry, a sector in which measures of preservation and revival are outstandingly costly.

These various considerations imply, the investigator found, that in most cases conservation interests will have to make their own financial arrangements by means of public funding, through voluntary contributions, and

Borås: the New Textile Museum is established in an old cotton spinning-factory from the turn of the century.
Photo: Borås Textile Museum.

Borås: demonstration of an old Jacquard loom in the Textile Museum.

Photo: Borås Textile Museum.

by closer involvement of business enterprises, this latter largely without the allurement of any immediate commercial incentives. The investigation has not yet resulted in any policy decisions.

Preservation Measures

The adapted reuse of old industrial facilities has continued, and special mention can here be made of Kalmar Ångkvarn (a former steam-driven corn mill), dating from the late 19th century, which has been turned into a county museum. The big dockside warehouses in Sundsvall and a textile mill in Norrköping similarly have taken on a new lease of life as museums. Several other examples can be quoted of factory buildings that have been converted into offices and homes. Restoration and preservation of mining environments are in progress, not least under the aegis of the ecomuseums that have been established in the past few years.

Borås consolidated its position as the textile center of Sweden during the 20th century, hence the establishment there of a textile museum. The museum occupies what used to be a cotton-spinning mill. The building is interesting in its own right, architecturally, and in the history of technology. Its general layout is also ideal for a museum. The museum plans comprise about 6,500 square meters on three stories. The basement is being used for archives, storage facilities, and workshops, while the big spinning rooms at ground level and on the upper story will form the basis of public activities.

The Sala silver mine workings date back to the 16th century. The surrounding countryside was transformed by an extensive network of dams. The mine, the human settlement, the dam and water supply system, and the various places with relics of mining operations are a significant part of the national cultural heritage. The historical development of the area has been analyzed, making it possible to put forward proposals to reconstruct previous working environments.

Existing and Planned Museums

All over Sweden are numerous industrial museums. Attention will be confined here to a number of projects now planned—again just a selection of the many and varied activities that are contemplated. Many of these projects can be classed as "cultural tourism." As a matter of course, every mining and manufacturing community and every old industrial center must retain buildings bearing witness to past ages, so that the history of the locality will not be forgotten. To develop into a genuine tourist attraction, the landmark must be unique in character and capable of offering something that is not available elsewhere.

Under the aegis of Ekomuseum Bergslagen, measures are being taken for the preservation and educational development of several attractions. In Västerbotten, in the far north a museum of mining and minerals is being planned. A broad, integrated approach to geology and mining history is intended to convey a knowledge of the evolution and historical significance of the northern mining industry. Mining is still being conducted, and the occurrence of sulphide ore in the Skellefteå

Järnriket Gästrikland: the "Iron Area" in the county of Gästrikland consists of half a dozen monuments and sites connected to the production of iron. The picture shows the restoration of the Jädraås Ironworks.

Photo: Tom Sandstedt, Gästrikland County Museum.

mining region makes the County of Västerbotten the largest nonferrous mining region in the country. The museum is to be integrated in a wider context as part of the Västerbotten Industrial Museum, which will also include a mineral center, for example. A national drilling core archives will be set up. Järnriket Gästrikland is a project for marketing the Gästrikland iron industry. The "Järnriket" project is marketing the iron industry of the province of Gästrikland as a common attraction. Industrial history, together with modern and earlier technical innovations, is presented in a holistic, humanistic perspective.

Summary

Space has not permitted an exhaustive account, and many interesting projects and schemes in progress have been omitted. The centralist perspective presented by way of introduction should not obscure the fact that the big efforts are to be found at regional and local levels. The new thing today is that legislation is being made to focus on targets. Legislation is an implement for bringing about the protection and preservation aimed for. Most important and hope-inspiring for the future, of course, are the energy and the profound commitment apparent from all agencies. Without those qualities, no targets can be achieved. This is a truism. The solution of the financial problems will be crucial.

Marie Nisser is author of the first part of the preceding chapter. Kersti Morger is author of the section beginning "Planning and Legislation."

Further Reading

Nils Björkenstam, *Västeuropeisk järnframställning under medeltiden (*with a summary in English), Jernkontorets Bergshistoriska skriftserie nr 26 (Stockholm, 1990).

Lasse Brunsström, *Kiruna-ett samhällsbygge i sekelskiftets Sverige*, with an extensive English summary on "Kiruna—A Swedish Mining City from the Turn of the Century" (Umeå, 1981).

Sherban Cantacuzino, ed., *Architectural Conservation in Europe* (London: The Architectural Press, 1975), pp. 77-83.

Helen Clarke, trans. and ed., *Iron and Man in Prehistoric Sweden* (Stockholm: Jernkontoret, 1979).

Ibid., Gert Magnusson, "Lapphyttan: An Example of Medieval Iron Production,"

E.F. Heckscher, *An Economic History of Sweden* (Harvard Economic Studies, 95: Cambridge, Mass., 1954; repr., 1963).

Karl-Gustaf Hildebrand, *Swedish Iron. Seventeenth and Eighteenth Centuries: Export Industry before Industratialisation*, Jernkontorets Bergshistoriska skriftserie (Stockholm, 1992).

Bengt Holtze, Åke Nisbeth, Rolf Adamson, and Marie Nisser, eds., *Swedish Industrial Archaeology: Engelsberg Ironworks—A Pilot Project* (Stockholm: The Swedish Royal Academy of Letters, History, and Antiquities, 1975).

Matts Isacson and Lars Magnusson, *Proto-industrialisation in Scandinavia: Craft Skills in the Industrial Revolution* (Leamington Spa, Hamburg, New York: Berg Publishers Ltd., 1987).

Marie Nisser, "Swedish Industry and its Monuments," *Transactions of the First International Congress on the Conservation of Industrial Monuments, 1973* (Ironbridge, 1975).

Marie Nisser, ed., *The Industrial Heritage. The Third International Conference on the Conservation of Industrial Monuments. Transactions*, 3 ed. (Stockholm, 1981).

Marie Nisser and Nils Björkenstam, *Ironworks and Iron Monuments: Study, Conservation and Adaptive Use*, Symposium in Ironbridge, 1984 (Rome: ICCROM, 1985).

Marie Nisser and Fredric Bedoire, "Industrial Heritage '84. Vol.1. National Reports 1981–1984," in Stephen Victor and Helena Wright, eds., *The Fifth International Conference on the Conservation of the Industrial Heritage* (Washington, D.C.: Society for Industrial Archaeology, 1984).

Protection of the Architectural Heritage of Sweden, Riksantikvarieämbetet (Central Office of National Antiquities) (Stockholm: The Architectural Heritage Year, 1975).

Kurt Samuelsson, *From Great Power to Welfare State* (Stockholm, 1968).

Bergslagen—the Iron Production Region in Sweden

Kersti Morger

For more than a thousand years, mining has left its imprint on the landscape and settlement of the Bergslagen region. Ever since prehistoric times, people here have been occupied with the extraction of ore and the production from it of gold, silver, copper, iron, and steel. This development can be followed in time and space, from prehistoric production sites to the modern industrial facilities of our own day. Few other places can spur the imagination like Bergslagen or offer such an abundance of visual knowledge. Bergslagen conveys an insight into how Sweden's industrial prosperity was founded and developed. Over long periods, copper and iron have been the country's main export commodities. Mining, smelting, and manufacturing have affected living conditions and the social fabric. They have also generated special patterns of settlement, still detectable from abandoned mine workings, the ruins of prehistoric and medieval metal-manufacturing sites, surviving furnaces and hammer forges, mining villages and manufacturing communities, Bessemer and open-hearth steel works, workers' housing from the days of the ingot steel processes, tenement buildings and pit wheels from the last expansive phase of the central Swedish mining industry.

The selection of Bergslagen as one large preservation area is not new. Already at the turn of the century, similar ideas were being broached, and since then, they have been discussed on many occasions. They were revived, for example, during the European Architectural Heritage Year in 1975. The past few years have brought a change in the objectives of architectural heritage management, with in-

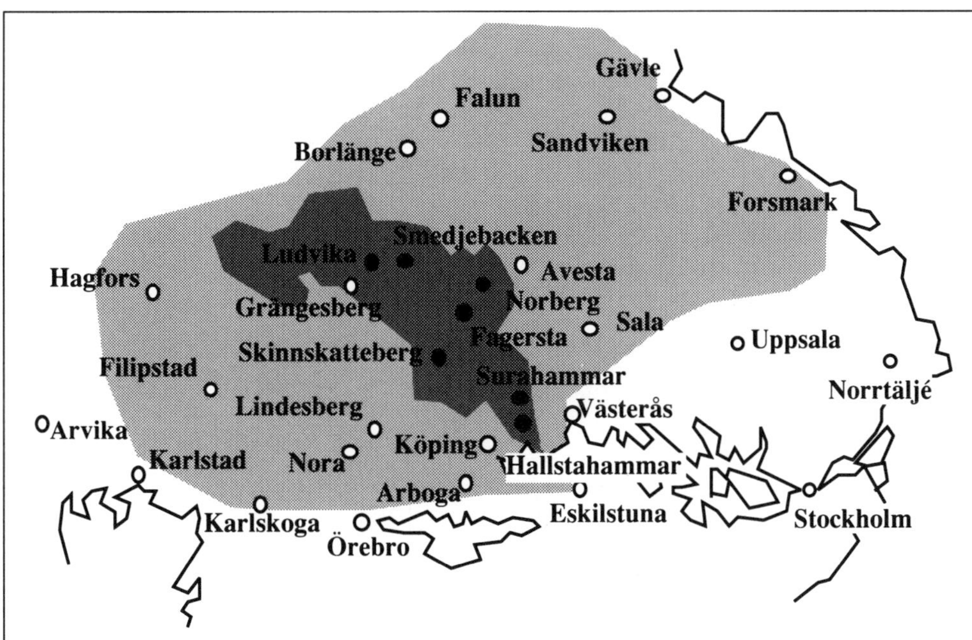

Bergslagen: Central Sweden, light grey tone is the Bergslagen area. The dark grey shows the Ekomuseum Bergslagen.

Bergslagen: Pershyttan blast furnace, just outside Nora, founded during the Middle Ages and enlarged during the 17th century. The present furnaces are from 1856. Production ceased in the 1950s.

Photo: Örebro Provincial Museum.

creasing emphasis on the preservation of complete environments. The Swedish Central Board of National Antiquities has specified a number of environments of national interest for future architectural heritage conservation. They include many industrial environments, several of which are located in Bergslagen.

The Bergslagen Concept

The principal mining and metal manufacturing areas are located in central Sweden and collectively known as Bergslagen. It was not till the end of the 19th century that iron began to be extracted on any considerable scale from the mines of Lappland.

The term Bergslagen has been given many different interpretations, and the same goes for its geographical extent. That discussion cannot be enlarged on in the present context, and only a few essential data of development will be presented here. Since copper production at the Falun Copper Mine is dealt with in another article, we shall confine ourselves here to a brief historical description of the iron industry. The Sala silver mine, admittedly, has played an important part, and its history is also worth describing, but this will have to be left aside for the moment.

History in Brief

Many of the low-technology iron production sites have been archaeologically investigated and datings of the oldest sites point to a relatively unanimous initial phase of iron production, viz, about 500 B.C. Iron production appears to have spread rapidly over southern Sweden, but some centuries were to pass before iron began to be made in Bergslagen.

Lapphyttan, a furnace site in the municipality of Norberg, was excavated in the 1980s. Using modern methods of archaeological analysis, it has been possible to date the blast furnace and the surrounding ruins. The results

Bergslagen: the Wallon forge at Österby bruk, the only one preserved with full equipment. Österby bruk is one of the oldest ironworks in northern Upland, within the Dannemora Bergslag, with iron production dating from the 15th century.

Bergslagen: Engelsbergs Ironworks is an extremely complete industrial site with most of the buildings preserved as well as the equipment. The oldest parts, the timber-clad blast furnace and the manor house *(top)*, are from the 18th century; the water-driven bar iron forge *(bottom)*, the workers' dwellings, and offices, etc. from the middle of the 19th century. Photos in STM.

of this investigation are sensational. The furnace site investigated consists primarily of a 14th-century structure, but there are strata indicating an earlier phase of iron production. All the indications are that the blast furnace was in operation in 1200 plus or minus 100 years. Presumably iron was first made at Lapphyttan some time in the late 12th century. We also know that the furnace was abandoned some time in the 14th century. Lapphyttan, then, would seem to be the oldest blast furnace ever excavated in Europe.

During the medieval period, the iron industry was conducted by the *bergsman* class, i.e., mine-owning farmers who occupied themselves with mining and iron production

Bergslagen: Dannemora Mines. The engine house for the Newcomen steam engine, in operation 1731-1735 for water-pumping work at the Norra Silfbergsgruvan.

during that period of the year when their attention was not preempted by agriculture and forestry work. The iron industry, then, was a seasonal enterprise. The *bergsmän* produced osmund iron for export. A great deal of this iron was taken to the ports on the other side of the Baltic. A flourishing wrought iron industry had developed round about these towns, converting the osmunds into bar iron for reexport to Western Europe. At the beginning of the 16th century, bar iron fetched almost double the price of osmund iron. For this reason, Gustav Vasa tried to introduce bar iron production into Sweden, but this was opposed by the *bergsmän*, and the king had a number of crown forges established, where bar iron was produced in water-powered hammer forges.

The 17th century was a period of rapid development in Swedish iron-making. Osmund forging was prohibited, and bar iron forging encouraged. Great advances were made in blast furnace and wrought-iron technology. New furnaces and hammer forges were started and output multiplied several times over. Sweden was now to be the leading European exporter of iron for a considerable period.

At this time, new entrepreneurs appeared who had a different social background from the *bergsmän*, namely the *bruk* proprietors. These included foreign financiers, members of the Swedish aristocracy, and people from the urban middle class. They had more capital behind them and were able to invest in larger facilities—especially bar iron forges—with higher output and longer operating periods. State mining policy was aimed at spreading the iron industry over wider areas than previously. To prevent a timber shortage, the heartlands of the iron industry in Bergslagen were to be reserved for the *bergsmän*—for their mining operations and their production of pig iron. Establishment of new bruk enterprises was encouraged in the surrounding areas.

The *bergsmän* went on blowing pig iron and also managed to retain a small share of wrought-iron production. They sold their pig iron to bruk enterprises, but many of the latter, especially in Uppland, produced both pig iron and bar iron.

Developments during the first half of the 17th century coincided with a period of great wars in Europe, especially the Thirty Years' War. Holland became a major customer, above all for armaments. After the restoration of more peaceful conditions, during the 1660s and 1670s, England became the principal importer of Swedish iron. The *bergsmän* furnace was generally run by several *bergsmän*. At the beginning of the period, the *bergsmän* accounted for practically all pig iron output, and the end of the 18th century they were still providing about half of it. One notable characteristic of the iron industry this time was its connection with the agrarian community. It has been said that the *bergsmän* were industrially active farmers while the bruk proprietors were industrially active landowners.

Sweden's iron industry was almost exclusively geared to exports right down to the second half of the 19th century. This in turn made Sweden closely dependent on developments in the world iron trade.

In quantitative terms, however, the individual bruk enterprises had a small output capacity. Total bar iron production at the Swedish ironworks, however, rose between 1814 and 1914 from 50,000 tons to 400,000. This steep rise in production was brought about by a combination of technical changes. In Bergslagen, a number of ironworks changed from the German forge technique to the Lancashire process, which combined with rolling mills made possible a radical growth of output. The smaller mills, being unsuitable for this type of production, were closed down. Further closures came with the introduction of the ingot steel processes, and a heavy concentration of enterprise became apparent, at the same time as the surviving units increased their output very quickly. The revolution in communications was a precondition of this transformation. The shakedown into fewer and larger units has continued ever since.

Bergslagen—an Important Part of Sweden's Cultural Heritage

It was in Bergslagen that Sweden laid the foundations of its industrial prosperity. Even today, steel is still being made in bruk communities with centuries-old traditions. The entire Bergslagen region can be seen as an immense ecomuseum with a wide span from low-technology iron production sites via medieval and more recent historical sites to new, modern steel mills. Together they convey a picture of how mining and metal production have shaped both landscape and settlement over the centuries. Even today there are sufficient facilities extant to illustrate the different phases in this development process. And yet the surviving historic environments of the mining industry are only a fraction of what used to exist.

Falun Copper Mine

Marie Nisser

Stora celebrated its seventh centenary in the summer of 1988. The climax of a long series of festivities, exhibitions, conferences, and seminars came with the annual general meeting of 16th June, when 6,000 shareholders gathered in Stora Stöten ("The Great Pit), the true genesis of the company's successful development over the centuries. It was on June 16, 1288 that Bishop Peter of Västerås signed a deed of exchange confirming that the diocese had acquired one-eighth of the mine, Kopparberget—"the Copper Mountain." Stora takes this as the starting point of its history.

In fact there is other evidence to show that people have been extracting the desirable copper ore and probably silver ore too from this mine for even longer. The likely origin of mining is still a topic for discussion, and in the past few decades, archaeologists, using new methods (C-14 analysis, pollen analysis, and spectral analysis of metals) have begun moving the initial date of mining operations at Kopparberg further back in time. At the beginning of the 1970s, the probable commencement of mining was dated at about 1000 A.D. Since then studies have, for example, been made of sedimentary deposits in the shallow Lake Tisken near the mine, and researchers have shown, more or less conclusively, that the metal content of these deposits changed considerably and became more and more concentrated during the 8th century, which could mean that mining and the smelting of metals were already being practised at that time.

To begin with, though, mining appears to have been almost an agrarian, common pursuit, on a negligible scale. But during the 13th century, Kopparberget had developed into a significant and viable industrial enterprise, possibly under German influence. Raw copper had begun to be an important export com-

Falun Copper Mine at the close of the 18th century. From a colored engraving by J.F. Martin.

modity. Shares in the mine were eagerly sought after, not least by Crown and church. The 1288 exchange was confirmed by King Magnus Ladulås (Lock-the-Barn), the archbishop, and five other members of the episcopacy. And at the beginning of the 14th century, we find the king of Sweden himself owning shares in the mine.

A succession of disastrous collapses during the early 16th century was followed by a drastic fall in production. Gustav Vasa took vigorous action to raise output again. German miners were called to Stora Kopparberget to construct hoisting devices for the ore, pump installations, and other facilities, and to improve the efficiency of mining operations. New and richer seams were discovered in about 1570. The new ore deposits, coupled with improvements in mining technology and work organization, opened the way to the great expansion that now ensued. Production in about 1600 was barely 450 tons, but output for the century as a whole averaged 1,500 tons. The figure for the peak year 1650 was twice that. Probably 3,000 tons of raw copper equalled two-thirds of European copper production at that time. Sweden had almost a

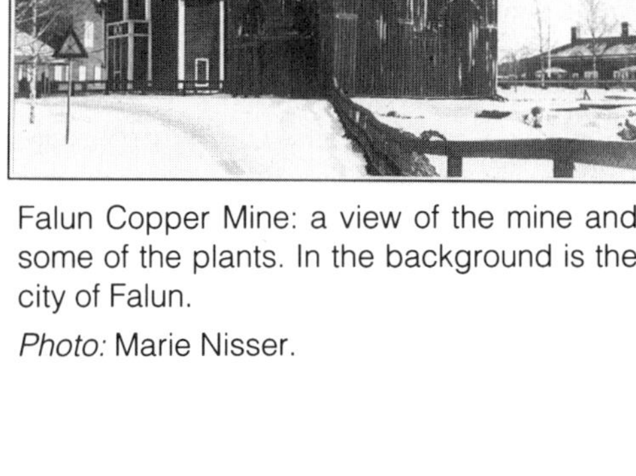

Falun Copper Mine: a view of the mine and some of the plants. In the background is the city of Falun.

Photo: Marie Nisser.

Falun Copper Mine: the Creutz shaft head.
Photo: Marie Nisser.

monopoly of the European copper trade. Mining operations were accelerated, vertical shafts had to be driven down to the ore, and the mine came to be increasingly undermined. There were heavy collapses. In 1655 the roof of the southern part of the mine collapsed, and a new open cast was formed. Just over 30 years later, in 1687, came the biggest disaster in the history of the mine, when the walls between three large open cuts collapsed, transforming them into one vast pit, Storstöten, the bottom of which was covered by debris. From 1691 onwards, production never exceeded 1,500 tons, and it fell off still more in the 18th century. Mining operations continued regardless, but Swedish copper could no longer hold its own in the world market. It was still Sweden's prime export commodity, second to iron, until about 1780, but this was followed by a heavy decline. The drop in copper output meant the inauguration of new lines of business. Production of sulphur and vitriol had been in abeyance during the heyday of the mine, but was now revived. Falun red paint began to be produced on an industrial scale in 1764, which provided a use for the copper-free iron sulphide. At the same time, a new charter was issued for a silver mine that had been idle since the 16th century. The mining company now also began systematically buying up shares in iron mines and ironworks, until eventually it became one of Sweden's leading enterprises in this industry.

The mine has been the starting point of the industrial activities that have evolved here, and that have invariably been characterized by change and development. Stora today is a world-wide enterprise. Copper ore is still being mined, and the iron sulphide from the mine is the raw material base of a profitable chemical concern. Falun red paint has a faithful circle of customers, and many people say it is unbeatable as a paint for timber buildings in the countryside and in wooden towns. Gold ore has been extracted in recent years.

Stora has wound up its iron and steel business and is now one of the leading European forest industry concerns.

Industrial Landmark Conservation at the Falun Copper Mine

Already in the 1890s, the then assistant manager of Stora Kopparbergs Bergslags Aktiebolag, Carl Sahlin, began putting the company's mining history collections in order. He was also destined to make a pioneering contribution to early Swedish industrial heritage conservation, an achievement that is one of the earliest of its kind internationally.

The 18th-century Gruvstugan, not far from Stora Stöten, was restored between 1917 and 1921, opening shortly afterwards as the mining museum. Continuing mining operations made it necessary to move the old buildings, including the museum, from the brink of Stora Stöten.

The older parts of the mine have been reinstated and are now open to visitors. The constricted adits and the mineshaft with its impressive timber structures, bear witness, in a centuries-long perspective, to the dangerous occupation of the mine workers and to almost incomprehensible drudgery in primitive conditions. Stora Stöten with its surviving headframes, waterwheel, and other buildings from various periods in the mine's history, makes up a holistic environment, one of the most concentrated industrial landscapes in Sweden.

Over the years the company has invested a great deal of money in conservation, restoration, and maintenance of other industrial environments belonging to it, many of which occupy a prominent position in the national industrial heritage.

Stora Kopparberget was the foremost European producer of copper during the 17th century. Its history can be traced back to the late 13th century in written records. Archaeological remains confirm the existence of mining operations at a much earlier date. The industrial landscape round about the Falun Copper Mine, with its great diversity of buildings from several centuries, the mine itself with its well-preserved original structures in the older shafts, and the mining museum together convey an evocative picture of a raw material base that played an important part in early European industrial history.

The Göta Canal

Bengt Nyström

The Göta Canal is a peaceful waterway through Sweden, reaching from the east coast to the west coast by way of various lakes and cutting through dry land. It was constructed with great visions of its future importance as a transport route, for industrial development and for military purposes.

The canal did not turn out to be the moneyspinner it had hoped to become. The railways put it out of business. But as a freight route and tourist attraction it has meant a great deal through the years.

Today, above all, the Göta Canal is a strong tourist attraction, but it is also a historic land-

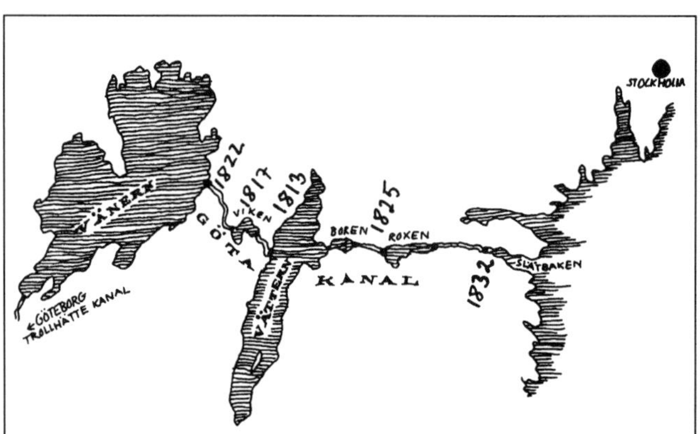

The Göta Canal: from Stockholm you can reach the canal in the narrow bay of Slätbaken by the Baltic Sea. The canal starts at the locks of Mem and stretches through Lakes Roxen and Boren up to Lake Vättern, over to Lake Viken and to the Lake Vänern. From Lake Vänern the canal continues through Trollhättan on the Göta River down to Gothenburg.

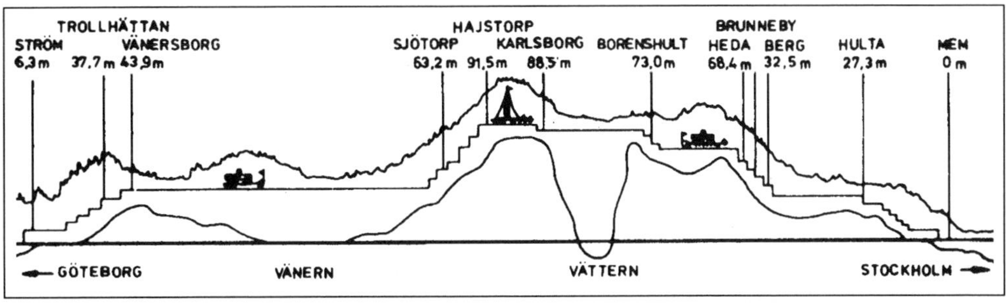

The Göta Canal: a profile of the canal showing the height of the locks. The highest point in Hajstorp is on 91.5 meters above sea level at Mem.

mark, in its entirety, in its details, and because of how it has affected social developments along its banks. It is also an outstanding industrial landmark, by virtue of its genesis, the technology that was used and the importance that it after all acquired as a vehicle for new technology and for incentives for technical progress in Sweden.

The first ideas for a navigation route traversing Sweden were already being discussed in the late sixteenth century. They were revived by several monarchs during the seventeenth and eighteenth centuries. The idea, in other words, was alive all the time, but the question took on a new urgency after Finland had been lost to Russia in 1809. Several factors helped to bring about this resurgence of interest.

The Example of Britain

During the second half of the 18th century, a network of canals had been constructed in Britain, mainly to meet the inland transport needs—raw materials, coal, and finished products—of an expanding industrial sector. There were, above all, two major projects that attracted attention on account of a series of technically interesting solutions and also because they meant so much to transport and industry: the gigantic Ellsmere Canal built in the 1790s, and the Caledonian Canal across Scotland, completed in 1801. The latter was also strategically important, providing the Royal Navy with a shortcut from the North Sea to the Irish Sea.

Similar visions were at work in Sweden. The construction of this great canal across the country would confer strategic advantages, making it possible for ships, munitions, and manpower to be transferred without difficulty from coast to coast, undisrupted by hostile fleets in places like Öresund—the Sound between Sweden and Denmark. Forces could be

The Göta Canal: the locks at Hovetorp in Östergötland, 1914.

Photo: Östergötland Provincial Museum, Linköping.

The Göta Canal: the locks at Mariehov, northwest of Söderköping.
Photo by: Jan Norrman 1991, the National Board of Antiquities.

rapidly concentrated to protect inland regions, in keeping with the doctrine of central defence that was being broached at this time and was part of the reason for building the fortress of Karlsborg, overlooking Lake Vättern.

The canal was also intended to shorten and cheapen passenger and goods transport operations, and some believed that it might even simplify transport operations from Russia to Britain and America. For example, it would obviate the heavy tolls that Denmark imposed on all ships passing through the Sound.

The prime purpose of the canal project was to encourage industrial growth and urban development alongside it. Britain was the main source of inspiration in this respect. A succession of vigorous industries would spring up parallel to "the blue ribbon of Sweden." Timber and agricultural produce would be exported from inland. Nothing came of all this. By the time the canal was finished in 1832, the railway age had already begun, and rail transport was soon to prove a more flexible system, more easily adapted to Sweden's industrial structure.

The Builder and His Engineer

The father of the Göta Canal was a naval officer, Baltzar von Platen. The Trollhättan Canal, linking Lake Vänern with Gothenburg, was finished in 1800, and the time was ripe to go ahead with the old plans for a waterway connecting Lake Vänern with Lake Vättern and the Baltic. Baltzar von Platen turned to the foremost of Britain's canal-builders, the architect and engineer Thomas Telford, who had been in charge of work on the Ellsmere

The Göta Canal. The Morala Machine Works in April, 1907. Established in 1822, the Works supported the building of the canal and the new steamers.

Source: The Canal Company Archives.

and Caledonian Canals. Telford was invited to Sweden in 1808, and that autumn, together with two of his engineers, he prepared a technical description of the canal from Sjötorp on Lake Vänern to Slätbaken on the Baltic Sea. Von Platen gained King Charles XIII's support for the project, and finally, in 1809, the Swedish Parliament granted a loan to the company that was to take charge of planning and constructing the canal.

Telford returned again in 1813 to site a number of locks and bridges over the canal, which he also designed. This time he was accompanied by several experienced lock-gate constructors and other craftsmen, who later took part in building the canal.

About ten regiments from Sweden's regular army contributed labor, and operations proceeded on a strictly military basis. Altogether the canal took 23 years to build, occupying a total workforce of something like 58,000. The work input has been estimated at 8 million day-works. The canal is 190 km long, including 87 km of artificial waterways. The 58 locks are individually named, and the heaviest concentrations of them occur at Sjötorp and Berg. This all makes the Göta Canal one of the biggest engineering projects ever undertaken in Sweden.

The First Stretch of the Canal, 1813

The first stretch of the Göta Canal to be opened to shipping was that between Lakes Viken and Vättern. The difference in altitude was overcome by means of a lock at Forsvik, which has the greatest lift on the canal. This stretch was officially opened with pomp and ceremony on 14th October 1813. The first boat went through the lock to music and the cheers of the onlookers. An infantry battalion paraded, the upper pair of gates and the drawbridge over the canal were opened, the name of the lock—Carl XIII—was proclaimed, and a nearby battery of guns fired a salute. The drawbridge, of cast iron, was the first of its kind in Sweden.

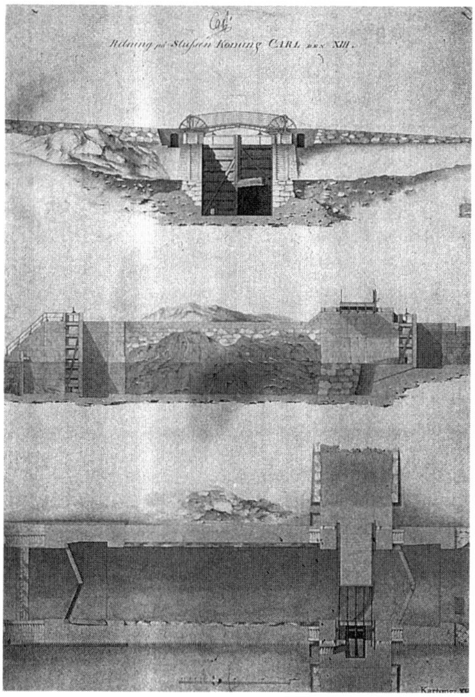

The Göta Canal: the locks and the drawbridge of King Carl XIII locks at Forsvik.

Courtesy: The Canal Company Archives.

The Västergötland section of the Göta Canal, linking Lakes Vänern and Vättern , was officially opened in 1822, after ten years' hard work. Motala Verkstad was founded that same year to assist the Canal builders. The Östergötland section was finally opened on 24th September 1832 by King Karl XIV Johan (Bernadotte), and the entire scheme was concluded between 1838 and 1844 with the modernization of the Trollhättan Canal.

New Technology and a Trade School

The building of the canal demanded a number of materials that were not to be had in Sweden: cast-iron lock gates, winches for them, steel structures for bridges, and other things besides. Most of all this was imported, mainly from England. Telford suggested setting up a special engineering workshop, and he offered to recruit workers and get things running. Motala Verkstad , completed in 1822, came to play an important, strategic part in the development of Sweden's engineering industry. The building of the canal needed a special material and repair workshop. Later production of steam engines and boats started. It was one of the first big engineering concerns, and it became a trade school for an initial generation of skilled workers and mechanical engineers. Its first principal was the Scottish mechanical engineer Daniel Fraser. The construction of the Göta Canal also provided a trade school for all the regular soldiers involved, who returned home bringing with them new technical knowledge to the rest of the country. The Göta Canal has in fact been dubbed Sweden's "first and longest institute of technology."

The Canal Company building in Motala, is now the home of a Canal and Shipping Museum. At Sjötorp there is an exhibition about the canal and the history of the Sjötorp Shipyard. Parts of Motala Verkstad are also to be preserved as an industrial museum, together with the lock gates and iron bridge at Forsvik, dating from 1813.

More Canals

The Göta Canal, though, was not the only canal project. Shipping routes have always been the main transport arteries for cargoes, and canal building was a logical development. In fact, the idea of canals was so implicitly believed in that, when plans were first made to construct railways, they were thought of, above all, as an adjunct to the canals. The Trollhättan Canal had already been opened in 1800, but it proved to be too small and was rebuilt between 1838 and 1844. The Hjälmaren Canal was widened between 1819 and 1839, and the Strömsholm Canal, opened in 1795, was enlarged between 1842 and 1860. The Dalsland Canal was constructed in the early 1860s, as were the Säffle and Kinda Canals. The Södertälje Canal, opened in 1819, became a vital link between Lake Mälaren and the Baltic, between Stockholm and "southern localities."

The Göta Canal: The Motala Machine Works, established in 1822. Lithograph by J. Jaeger from 1886. *Source:* the Nordic Museum Archives.

Frövifors Paper Mill

Helene Sjunnesson

Frövifors Bruk, 30 km north of Örebro, is a large, modern sulphate and board mill owned by ASSI, a state-owned forest industry company. Surrounded by this enormous, vibrantly busy industrial landscape, the old paper mill, built in 1902 on the upper waterfall of the Frövi river, still remains. Because the new board mill, commissioned in 1981, was built separately from the old paper mill, essential parts of the latter could be saved.

The idea of preserving the old paper mill was already broached in 1975, when the management first began planning a new cardboard mill at Frövifors. The initiative for the preservation came from the company head of public relations, Sven-Göran Wennerholm, who contacted the historical committee of the Swedish forest industries, set up in 1973. After an inspection of the site and subsequent documentations of pulp and paper mills in the several regions, it is now clear that the old Frövifors paper mill is the best suited plant for preservation in Sweden from the early era of machine-made paper. Scholars and antiquarians, industrial representatives, and unions all agree on this point. What is more, the location and industrial history of the mill are typical both of the industry and of Sweden's industrial development in general. A waterfall and surrounding forests were essential to the founding of an ironworks in the 16th century. The ironworks was forced out of business in 1890 as part of the structural transformation of Sweden's iron industry at that time.

Meanwhile, an emergent European paper market and new technology for the mass production of paper out of wood, laid the foundations of a new industrial enterprise in the locality. The surrounding forests, earlier used for charcoaling, provided raw material for the groundwood pulp and paper mill built in 1891. The waterfall supplied energy and process water. Housing was available on the spot, and the workers were taught new skills.

Frövifors Paper Mill: the paper mill in 1907.

Photo: the Swedish National Museum of Science and Technology.

Industrial Overview

It was at the time of Sweden's industrialization, during the second half of the 19th century, that the modern pulp and paper industry based on wood advanced to be one of Sweden's basic enterprises and a major export industry. Paper, it is true, had been manufactured here since the 16th century, but until the 1830s, it was produced one sheet at a time and by hand, using textile rags as raw material. Two technical prerequisites for large-scale production were the paper-making machine, which became industrially viable in the 1820s, and processes for manufacturing paper from wood-fibers which were devised between the 1850s and 1870s. Sweden being a heavily forested country, pulp mills were rapidly constructed, often in conjunction with mechanized paper mills. As a result, the numerous hand-made paper mills were gradually put out of business. One of the last to keep going was the small, antiquated Ösjöfors mill in the province of Småland, which closed down in 1926. That is the only one of the small hand-made paper mills still surviving *in situ*, and today it is a living museum, used in the summer to demonstrate the old manual craft of paper making.

During the 130 years it has now existed, the wood-fiber-based pulp and paper industry has undergone great changes. As in so many other industries, the trend has been away from numerous small factories with large work forces, to progressively fewer and larger processing plants owned by conglomerates, and with supervision and computerized control taking the place of machine-tending.

Why Frövifors?

There are some pulp mills and mechanized paper mills where only the buildings survive, the machinery having been sold or scrapped. The older facilities still in operation have been successively modernized and enlarged, so that as a rule, only parts of buildings or fragments of machinery bear witness to earlier periods in their history. It is unusual to find such an unchanged early paper mill as Frövifors.

The old mill area is reached by a stone bridge built in 1818 and past a manor house from the 18th century. Otherwise, though, a number of slag heaps are the sole relic of the iron-working era. Nor is there really anything left of the 1891 pulp and paper mill. The ground-wood mill was soon replaced by a power station, and a new paper mill and sulphate mill for producing craft paper were built in 1902. The sulphate mill is still working but has been completely rebuilt.

The interesting part of Frövifors, in terms of industrial history, is the redbrick machine

Frövifors Paper Mill: workers at a paper machine at Frövifors in the 1890s.
Photo: the Swedish National Museum of Science and Technology.

halls containing two original Yankee paper machines, called PM 3 and PM 4, and made by Karlstads Mekaniska Verkstad in 1907 and 1911 respectively. The first of these, especially PM 3 from 1907, has hardly changed since it was installed. Also of interest from the viewpoint of technical history are transmission shafts for the belt drive powered by a water turbine, used until 1942, as well as kollergangs and cast-iron hollanders for grinding the pulp—once a standard item of equipment at paper mills but now superseded by other devices.

The paper machines have been idle since 1976, when the mill was closed down. Running them on a museum basis does not seem, unfortunately, to be a realistic proposition. To see what things were like when the mill was working, one can study the film and photographic documentation compiled in 1976 just before the closedown. The mill also has comprehensive archive materials. There are Frövifors residents whose recollections can bring this silent environment to life again. Besides, what we have here is not just a unique "time capsule" from a vanished period, but a living industrial community, a meeting point of old and new. Visitors to the museum area can also enter a modern paper mill, while younger generations working at the mill have an opportunity of viewing their work in historical perspective.

If it had not been for the indefatigable efforts of enthusiasts at Frövifors, the preservation plans would probably have come to nothing, even though many people supported the project and joined in working parties and investigations, such as the historical committee of the Swedish forest industries, the Örebro County Museum, the Örebro State County Administration and County Council, the Central Board of National Antiquities, the National Museum of Science and Technology, the Municipality of Lindesberg, the Näsby local history association, and the Friends of Frövifors Paper Mill Museum. At their suggestion, in 1984, the Frövifors Paper Mill Museum Foundation was established, and a donation and leasehold agreement signed with ASSI.

Preservation Ambitions

Money, as usual, is the crunch. Large sums have been needed to preserve the building and machinery, but today, 16 years after the first ideas of preservation were put forward, the historic paper mill seems at last to have a bright future.

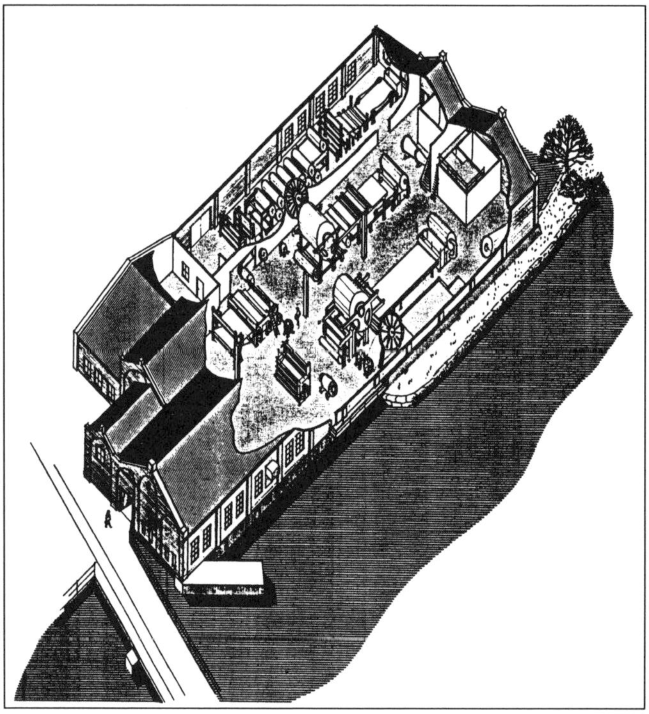

Frövifors Paper Mill: the machine hall in the old mill. Today there are two Yankee paper machines and a plan to set up a cardboard machine from the 1860s, received from another mill.

Drawings: Architect Hans Eriksson, Frövifors Paper Mill Museum.

The most necessary improvement—relaying the roofs of the machine rooms—was made possible in the winter of 1989 by grants from several sources, e.g., the National Labor Market Board, the Örebro County Council, the municipality of Lindesberg, the Swedish Paper Workers' Union, the Örebro State County Administration, and the Central Board of National Antiquities.

But the aims of the foundation extend further than preserving and demonstrating the old paper mill. To attract visitors to Frövifors, a paper industry museum is planned with exhibitions providing nationwide, regional, and local perspectives. Advanced plans have been made to this end. Exhibition premises, personnel facilities, and an archive depot are to be housed in a former repair shop near the river. The foundation has received some old machines from other mills. One of these is a cardboard making machine, made in Scotland in the 1860s, which until recently was used at Skogshallsverken. This is to be set up in a hall that is now being built next to the PM 3 hall. Other larger equipment will be placed outdoors. Cooperation with the Örebro County Museum is continually expanding. Two permanent exhibitions have recently been opened at the mill. One is about the history of Frövifors mill and the other one is entitled "Packaging From Stone Age to Space Age."

To realize the foundation's full plans will amount to some more million Swedish crowns. If the entire project is accomplished, the result may be a Swedish counterpart of the Klevfos Industrial Museum in Norway. Looked upon in a Nordic perspective—the development of the pulp and paper industry has followed similar patterns in Sweden, Norway, and Finland—there are good reasons for preserving the Frövifors paper mill. In Finland there are the well-preserved groundwood pulp and cardboard at Verla from the 1890s and the Sunnila chemical pulp mill from the 1930s. In Norway there is the sulphate mill and small paper mill at Klevfos from the early 20th century that closed between 1972 and 1976. If just one mill of this format can be saved, cared for, and kept alive in each country, quite a lot will have been achieved towards preserving the industrial heritage of a branch of such great importance to the Nordic countries. Things will be more difficult when the industries of today start becoming "history."

The Industrial City—Norrköping

Gunnar Sillén

Hardly any other place in Sweden can rival the powerful onrush of spring in Norrköping, when the sun has got the better of the snow and ice of winter, and the water from Lake Vättern and the Östergötland countryside surges through the body of the city on its way to the sea. Crossing the narrow footbridges over the ponds in between the factories on either side of the Motala River, one is almost bowled over by giddiness from the thunder and turmoil of water beneath one's feet.

The massive stonework of the Bergsbro bridge makes a slightly more solid vantage point in this peculiar riverside industrial landscape. The factory buildings themselves are quiet nowadays. In times gone by, and especially in the summer, when all the windows were open to let out the heat, they emitted a deafening noise—a whirring of transmission belts, the rattle of spinning machines, and the thump of power looms. Smoke billowed from the great chimneys, and morning, noon, and

Norrköping. The well-preserved industrial landscape along the Motala river.
Photo: Ingvar Anehed, Museum of Work, Norrköping.

Norrköping: Norrköping's cotton textile industry was built in 1856 close to the stream. Three water turbines transformed the energy. The weaving and spinning machinery was imported from England. Tohe Manor House (from 1829) to the right belongs to the adjacent factory: the Gryt Textile Mill / L J Söderberg & Arosenius.

Lithograph from *Sv Industriella Etablissementer*, 1872.

night, the day was punctuated by the screeching of factory whistles.

Old pictures of Norrköping frequently express pride in the factories and the industrial landscape. The belching chimneys had yet to be seen as pollution problems, the prosperous captains of industry as bloodsuckers. The regimented windows of the factory buildings had yet to become tedious. Even night work imparted a kind of beauty to the townscape, with the factories glowing in the darkness like illuminated castles.

The textile industry

Norrköping's industrial landscape is dominated by the textile industry, developments in which were founded on high mechanization and low wages. By the end of the 19th century, competition was already making it difficult to sustain both profitability and wages. Companies were amalgamated, and gigantic new, "rational" factories built to overcome these problems. Since, in practice, women were forced to accept more rigid control and lower rates of pay than men, as time went on, the workforce in the factory landscape became predominantly female.

The working conditions of women in the weaveries and spinneries of Norrköping did not exactly improve with the passing of time. True, the most dangerous machines were encapsulated eventually, to save lives and expense, but as machines grew faster and more numerous, the noise increased. With 600 looms in one room, not even shouting was audible. Time and motion studies and the whiplash of piece-rates compounded the grimness. Sitting on the job was not allowed. Running was dangerous because a slippery patch of oil could send one diving straight into the machinery. The air was thick with dust and water vapor.

The Drag Factory and the Central Prison

Looking upstream from our vantage point on the Bergsbro bridge, we can see, furthest away from us, the Drag factory with its two big gables overlooking the Motala River and the uppermost of its many weirs within the city. Christian Lenning began manufacturing woollens here on a large scale in 1810, but glovers, dyers, and broadcloth manufacturers had been upgrading wool on this spot since the 17th century. The factory's location next to the Central Prison is no coincidence. That blend of philanthropy and utilitarian thinking peculiar to the Industrial Revolution led to the construction of the prison's precursor, a spinnery, in 1790, as a kind of annex to the factory. The factory owner could rent cheap labor from the spinnery, and thus the prisoners (all women)

Norrköping: the "flatiron," the well-known profile of the industrial landscape on a small island in the middle of the stream. The cotton spinnery was built in 1917 in reinforced concrete for the Holmen Textile Mills, which had several adjacent factories, some dating from the 1880s. It has been newly renovated for the Museum of Work. In the background older factory buildings and the paper mill, which recently moved out. To the left the old factory buildings now used for the City Museum of Norrköping. *Photo:* Gunnar Sillén, 1975.

were graciously invited to become and feel like useful members of society.

Incorporation in 1854 laid the economic foundations of Drag's development into one of Sweden's leading woollen factories. The facilities were enlarged several times. Workers' homes were built in Hållhaken precinct. Lecture rooms and libraries were established for their betterment and the education of their children.

Drags closed down for good in 1954, by which time its machinery was worn out, profitability was wretched, and the workforce down to about 250. Since then, the buildings have been vacant or have been used as warehousing or applied to various temporary purposes. There are strong demands today for converting the place into a housing development.

The other former textile mills upstream of Bergsbro bridge include the Gryt and Berg cotton mills and the Ström and Bergsbro woollen factories.

Downstream of the bridge, on the left between the water and Västgötegatan, the scene is little short of idyllic. Here there used to be flour mills and medium-sized cotton mills, and in the middle of it all is a garden with a summer house built by one of the factory owners, right on the water's edge. Many of the old factory buildings have now been converted for use by the City Museum of Norrköping.

Holmens Bruk

Downstream from the Bergsbro bridge, the view is dominated by a building tailored to the western point of Laxholmen and commonly known as "the Flatiron." It cleaves the flowing water like the prow of a swift-moving ship. Today the Flatiron is being converted into the Museum of Labor, but it began life, in 1917, as a cotton weavery owned by Holmens AB. The yellow rendering and the classical architecture are meant to tell us that the building belongs with a number of other factories owned by Holmen and is part of Norrköping's earliest history.

Standing here a couple of centuries ago, on Bergsbro's predecessor, one would have seen a welter of small islands, ponds, flour mills, and buildings and engines of all kinds. In the distance, on Kvarnholm which at that time was still a separate little island, Holmens Bruk soared above the water and the muddle—a close-knit manufactory dominated by its gateway tower, constructed in 1751. Kvarnholmen ("The Mill Island") had originally been the home of a number of Crown factories (for small-arms production) that in 1627 were taken over by the Walloon industrialist Louis de Geer and converted into a brass foundry with various sidelines. Brass was a prestigious commodity but not always as profitable as, say, flour-milling and salmon fishing.

The mill was burned by the invading Russians in 1719 but rebuilt with several new operations added to it. Paper-making was started in 1806, cotton-manufacturing in 1854. It was above all paper-making that helped to a completely transform of the factory landscape in and around the stretch of river between the Bergsbro and Gammelbro bridges. To guarantee the paper mill the power supply and feedstock water it needed, Holmens Bruk steadily bought up all the available mills and installations with riparian rights in the area. Waterwheels and turbines were demolished, traditional occupations—salmon fishing, for example—were abolished. Kvarnholmen and Laxholmen were joined together. The river was straightened out to maximize the efficiency of the headrace for the new power station that in 1923 replaced all previous power plants on the Motala River downstream of Bergsbro bridge. If the Flatiron, erected a few years earlier, was the prow of Holmen's imperious manufacturing plant, then the Bergsbron-Havet power station was the stern—the point at which the water eddies away from the propellers of Holmen after relinquishing most of its power to turbine wheels and electricity generators.

The paper mill grew. To accommodate the growth, the river was partly filled in or built over. The lack of space and, not least, the heavy traffic congestion resulting from the daily necessity of admitting many overloaded timber lorries and their trailers to the heart of Norrköping, finally made it necessary to close down first the pulp manufacturing plant and then the paper mill. The rattle of the timber intake was silenced. And a fragrance disappeared together with the white clouds of steam from the paper-making machinery.

Factory Landscape a Listed Area

Before the 1960s, hardly anyone gave a thought to the potential historic interest of this Norrköping factory landscape. The authorities did their utmost to clear away all the buildings left vacant by closed factories and resettled workers. Everything recalling the wretched working conditions of the past and the economic decline of the present was to be obliterated. (Norrköping's industrial problems were indeed formidable: 86 companies with a combined workforce of 8,000 went out of business between 1951 and 1971.)

More and more people, however, had come to realize that this factory landscape is part of Norrköping's identity, and efforts have been made to devise activities that could infuse new life into the area without destroying its history. Several official enquiries have been mounted to define and protect the historical interest of the area. A strong body of opinion has been mobilized in the city to prevent the Holmens Bruk company from going ahead with its plans to amalgamate all hydropower and all weirs and waterfalls in Norrköping into one new, efficient dam and a single new power station. If that were to happen, the Motala River would be turned into a mill pond from which, at best, business interests would allow small quantities of "cosmetic water" to gush forth as public entertainment on special occasions.

The onward rush of water has been the life blood of Norrköping's history, in things both great and small. Skvallertorget Square, at one end of Bergsbro bridge, for example, is named after the most primitive form of water mill—the undershot "Viking" mill (skvalta). These were the mills that ground the grain from the rich plains of Östergötland, making it possible to feed all the workers needed to run the factories hereabouts. And it was the water that powered the factories. This was where copper and iron could be upgraded. Products from the Bergslagen mining communities could be exported in ships that eventually returned with cotton, new machinery, and the technical designers of a new age.

As the water rushes onward between the Drag and Gryt factories, under the Bergsbro bridge and past Holmen towards the harbor, one perceives that it is this water that, right here, at the point where it reaches the sea, has fertilized the encounter between iron and soil, engendering the city of Norrköping with all its buildings and revolving wheels.

Porjus Hydroelectric Power Station

Marie Nisser

Sweden, with its abundant resources of hydropower, was swift to exploit the development of electrical engineering. In the first half of the 1890s, a Swede, Jonas Wenström, was one of those who successfully transmitted power over long distances by using alternating current with several phases. This opened up new prospects of electrifying waterfalls and using the power for industrial purposes, as well as transmitting it to remote urban and rural communities. Power stations immediately began to be built all over the country. One of the first of them, Näs power station in Dalarna, built between 1896 and 1899, has been preserved as a museum. During the early 20th century, the Swedish government became increasingly involved in national hydropower development. Through the State Power Board, set up in 1908, the government was responsible for the construction of Sweden's three most important power stations: Oliden at Trollhättan, opened in 1909; Älvkarleby on the Dala River, commissioned in 1914; and the great hydropower station at Porjus, on the Lule River, completed in 1914. Of these three power stations, each of them among the most significant monuments in the history of Sweden's power production, we have chosen to present Porjus, "the Temple in the Wilderness."

Porjus took four years to build. Parliament approved the plans for it in 1910, after which there ensued an immense construction enterprise that has been described as "a victorious campaign against the wilderness." Porjus is in the far north of Sweden, a few miles north of the Arctic Circle, downstream of the gigantic storage reservoirs of Stora Lulevatten, from where the Great Lule River descends in a sequence of rapids. At the time this wilderness was inhabited only by a few Lapps. The nearest village was 50 km off, and there was no road to it. This was where a new community was to be built up, and this was to be the site of the world's first major power station with underground machine rooms—all of which was to be accomplished in adverse climatic conditions that included snow, ice, and severe cold during the winter. A railway was planned between Gällivare and Porjus to bring up supplies of provisions and construc-

Porjus Power Station: the hydroelectric station was built in 1910-1915.
Photo: The National Water Power Archives.

tion materials. The railway and power station were built simultaneously. During the early months, everything coming to Porjus had to be carried by strong men walking on duckboards over squelching wetlands. This punishing journey took a day to complete, with each man carrying between 45 and 70 kg. There was another, less arduous route along which cargoes could be carried on boats for some stretches and towed by horses. That journey took three days.

None of the 20 or so power stations built previously in Sweden involved anything like the number of operations and vicissitudes as Porjus. A 1,250-meter dam had to be built to increase the drop and even out the supply of water. The whole construction was designed with the cold climate in mind. The dam would have to withstand the tremendous pressure of the ice. A 525-meter headrace tunnel was blasted into the rock to keep the water from freezing. That tunnel opened out into a dividing reservoir with five tube inlets. From each of these inlets, the water plunged 50 meters through tubes encased in concrete and sheet metal.

The machine room, then, was 50 meters below ground. Its proportions were impressive: 70 meters long, 11 meters wide and 11 meters high. At that time Porjus was probably the biggest underground power station in the world.

From the generators in the machine room, power cables were taken through a special shaft up to the switchyard building, which included transformers, a control room, and a high voltage room. The design of this station included provision for a future expansion. In its initial state, five turbines were constructed with a combined output of 38 MW.

The switchyard building was a monumental structure designed by Erik Josephson, one of the foremost architects in Sweden. He also designed the Trollhättan and Älvkarleby power stations. The building, of warm red brick, rests on a base of rough-hewn granite blocks. The entrance, at one end of the façade, is marked by a projection above which there rises a tower, surmounted by a royal crown. Porjus was to be a power center, a generator of industrial development in the north and a source of cheap power for the railway going from Luleå, through the Lappland mine fields of Gällivare and Kiruna, to Riksgränsen on the Norwegian border and from there to the port of Narvik. But the industrialization of Porjus did not live up to expectation. The electrification of the railway, on the other hand, was a success, which in turn provided a crucial stimulus for the ongoing electrifica-

Porjus Power Station: from the building of one of the feeder stream canals to the turbine hall.

Photo: The National Water Power Archives.

tion of Sweden's rail network. In the village of Porjus, the hutments of the construction period were superseded by architect-designed villas for the white-collar staff and housing for the workers. Porjus expanded, its population peaking at 2,000-3,000 between 1920 and 1950.

For more than 35 years, practically all Norbotten's electricity supply came from Porjus. Following an expansion in the 1970s, Porjus is one of the biggest hydropower stations in Sweden. Today it has a generator capacity of 530 MW—more than ten times the rating after its first enlargement in 1915, and a population of about 600. Vattenfall (the State Power Administration), with a personnel strength of 170, is its biggest employer.

A large operational control center opened in Porjus in 1988, now supervises and controls five power stations on the Luleå River. This is not the end of the Porjus story. The monumental switchyard building was threatened with demolition in the early 1970s. A new underground station was to be built, and the State Power Administration saw no reason to keep up a redundant building. The demolition threat prompted a widespread reaction. The heritage conservation authorities visited Porjus, and shortly afterwards it was proposed that the switchyard should be listed as a historic building, to save it from demolition. At the same time the Archives Committee, the active association that cherishes the history of Porjus, started canvassing public opinion and lobbying politicians to convince them of the necessity of preserving Porjus. In a letter to the Prime Minister, they asked: "Is not the history of the workers also culture? Or are we to go on believing that Gustavus Adolphus and Charles XII are the only ones that have meant anything to Sweden?" A strong body of opinion emerged. Vattenfall announced a competition for schemes to reuse the building. Alf Wallbing, a Porjus hydropower engineer, put forward an integral scheme of ideas, contributed by many people in Porjus, for turning the switchyard into a center of manifold activities. The crucial decision was made by Jonas Norrby, at that time Director General of Vattenfall. He came out in favor of the building being preserved, and that is what happened. The enormous brick building was converted into workshops, offices, and personnel facilities of high quality. A large, modern indoor gymnasium was fitted out on the third storey. The vigorous Archives Committee was allocated premises for exhibitions, for archives, and for its tireless and enthusiastic efforts to document the history of Porjus. The marble panelling of the control room has been preserved. Something like 20,000 tourists come to Porjus every year, mostly during the summer. In the switchyard building, they can visit the exhibitions and take a break in Lina's Cafe, now resurrected as one of the many cafes of the pioneering age.

The bill for all this was a heavy one, but the investment paid off handsomely in the form of practical facilities. The switchyard building, with its excellent workshop facilities, exhibition space, and gymnasium, has come to be a center of work, leisure, and tourism in Porjus.

At one time the Temple in the Wilderness was a symbol of great visions of the future. The Minister for Home Affairs put those expectations into words at the official opening of Porjus on 8th February 1915: "We see here the consummation of a gigantic labor that, in the midst of the wilderness, rises like a majestic monument to Swedish enterprise, to Swedish technical ingenuity, to Swedish valor. It stands here also as tangible proof that one of the grand dreams that have been dreamt about the future of Norrland has become a living reality."

Porjus today has a different vision of the future. Today the switchyard building symbolizes a community that has become an outpost a few miles north of the Arctic Circle—a community that was engendered by the building of the power station, a community whose existence hinges on the power station's survival. To the people of Porjus, the campaign to rescue the switchyard building meant a great deal more than averting the demolition of an important architectural and industrial landmark. What they were fighting for was the continuing existence of their community.

Notes on Contributors

BJØRN EDVARDSEN, manager, was trained and has worked as a teacher. He also worked as project leader for establishing the Industrial Workers´ Museum in Vemork–Rjukan. The museum is situated in a redundant industrial building that in 1911 housed the world´s largest hydroelectrical power plant. He is now manager of the museum.

ELIAS HÄRÖ is an art historian and since 1988 has been Department Director at the Finnish National Board of Antiquities and Historical Monuments. He has published several articles on the industrial heritage and the preservation of historical monuments. He is chairman of the Finnish Industrial Heritage Association.

JAN HULT graduated from the Royal Institute of Technology in 1950 and in 1957 received an Sc.D. at MIT. He divides his time between applied mechanics and the history of technology as professor at Chalmers University, Gothenburg. His main research interest is the strength of materials at high temperatures. Since 1988 he has been chairman of the Swedish National Committee for the History of Technology and coauthor of *Svensk teknikhistoria* (A Swedish History of Technology), published in 1989. Since its foundation in 1983, he has edited *Polhem*, the Swedish quarterly journal in the history of technology.

OLE HYLDTOFT, Dr.Phil., is Associate Professor at the Institute of Economic History at the University of Copenhagen. His main interests are industrial history, the history of technology, and housing and urban history. He is currently working on a book about the technological history of Danish industry 1870–1930. His writings include *Københavns Industrialisering 1840–1914* (1984).

MANGE JERVAN, M. Sc. Eng, has been Director of the Klevfos Foundation since 1982 when it was founded. Before his present affiliation, he worked in Norwegian industry for 25 years.

SVANTE LINDQVIST is professor of the History of Technology at the Royal Institute of Technology, Stockholm. He holds a M.Sc Eng. (Physics) and a Ph.D. in the History of Science and Ideas. He has published *The Teaching of History of Technology in USA: A Critical Survey in 1978* (1981) and *Technology on Trial: The Introduction of Steam Power Technology into Sweden 1715–1736* (1984). He is coauthor of *Svensk teknikhistoria* (A Swedish History of Technology). In 1986–87, he was a visiting scholar at the University of California-Berkeley. He has been a member of the Executive Council and the Advisory Council of SHOT.

TORLEIF LINDTVEIT has been Director of the Norwegian Museum of Science and Technology since 1967. During this period the museum has moved to a new and larger building. He has also been extensively engaged in the preservation of industrial monuments on site, and in international museum work. Since 1986 he has been chairman of the International Committee of Museums of Science and Technology.

DANCKERT MONRAD-KROHN, Ph.D., is Regional Inspector of Monuments and Sites in Vest-Agder County in Norway. His interest is the industrial heritage, for instance, lighthouses and waterpowered sawmills. He is also takes interested in developing tourism based on the local heritage. He has worked with the preservation of waterpowered sawmills in western Norway and is currently working on a project to preserve and reuse lighthouses.

KERSTI MORGER, Ph.D (Human Geography), is Senior Executive Officer at the Swedish Central Board of National Antiquites. Her dissertation dealt with technology and social change at a Swedish ironworks during the 1870s, *Skebobruk: Teknisk och social förändring vid ett järnbruk under 1870-talet* (1985). She is now working on the industrial heritage in Sweden.

TIMO MYLLYNTAUS, Ph.D., is Senior Lecturer in Economic History and the History of Technology at the University of Helsinki. After earning his Ph.D. at the London School of Economics, he has researched and written on technology transfer and comparative economic history. He has also lectured on the history of technology at the Tampere University of Technology. His research interests are the history of energy use and the environment. He is the author of *Electrifying Finland: The Transfer of a New Technology into a Late Industrialising Economy* (1991).

GUNNAR NERHEIM is Lecturer in the History of Technology, Rogaland University Center, Stavanger. He has contributed to several books dealing with the history of Norwegian business and technology. His last contribution published in English is in Kristine Bruland (ed.), *Technology Transfer and Scandinavian Industrialisation* (1991). Presently he is studying the diffusion and development of offshore technology in Norway, to be published in a two-volume *History of Norwegian Oil* in late 1992.

HENRY NIELSEN is Associate Professor at the Institute of Physics, University of Aarhus, in Denmark. His research interests are shared between educational science and the history of science and technology. He is coauthor of the textbook *Skruen uden ende* (The Endless Spiral–A History of Western Technology) (1990).

MARIE NISSER has been appointed to a new chair for Industrial Heritage Studies and Research set up by the Swedish Council for Research in the Humanities and Social Sciences at the Royal Institute of Technology in collaboration with Uppsala University. Her main research field is the industrial heritage with special emphasis on iron and steel, paper and pulp, the textile industry and historic building and planning. She has been the chairperson of the International Committee for the Conservation of the Industrial Heritage (TICCIH) in 1984–90.

BENGT NYSTRÖM, Ph. D., is Vice-Director at the Nordic Museum in Stockholm. He is former Director of the National Museum of Science and Technology in Stockholm and of the City and Industrial Museum of Norrköping. He has done research in, and published several publications on, various museological topics, among them industrial archeology, crafts, and film. He is also editor of *Teknik & Kultur*, a journal for the history of technology, industry, and labor.

OLA ØVERÅS, Master of Architecture, is principal inspector at the Central Office of Historic Monuments and Sites. His interests are cultural history with special attention on the history of technology, industry, and communication. He is presently working on a Norwegian working history to be published in 1992.

LAURI PUTKONEN, B.A., is an art historian and has since 1970 worked in the field of building protection. From 1970–76 he worked as researcher at the Finnish National Board of Antiquities. In 1976 he became a private consultant and has carried out several regional and national inventories of historic monuments and sites. He has published various articles on architectural history, industrial architecture, and building technology. He is secretary of the Finnish Industrial Heritage Association and the national representative of Finland in TICCIH.

GUNNAR SILLÉN, architect, author, and artist, initiated with the book, *Stiga vi mot ljus* (1977), the documentation of industrial work and the preservation of industrial monuments as a task for study circles within the Workers Movement. His present project is a book on the interpretation of the aesthetics of industrial culture as it has been expressed in ideas, organization, technology, politics, work, and design.

HELENE SJUNNESSON, B.A., is curator at the Department of Documentation at the Swedish National Museum of Science and Technology. Her main interests are industrial history and the preservation of industrial monuments. She has published, together with Marie Nisser, a report on the pulp and paper industry, *Massafabriker och pappersbruk i Värmland och Dalsland* (1973). Presently she is working on a book on pulp and paper in southern Sweden, *Papper och massa i Östergötland.*

MICHAEL F. WAGNER is Research Fellow and editor of the Danish historical periodical, *Den Jyske Historiker*. From 1987 to 1991, he worked as Assistant Professor at the Technical University in Aalborg. Since 1991 he has been at the University of Aarhus, working on a major thesis on polytechnical environments in Denmark during the 19th century. He has written a number of articles on the history of technology.

HANS WEINBERGER, M.Sc. Eng. (Engineering Physics) is a graduate student at the Department of the History of Science and Technology at the Royal Institute of Technology, Stockholm. He has published a biography of the founder of Swedish radiophysics, radiobiology, and radiation protection, the physicist Rolf Sievert, *Sievert: enhet och mångfald* (Sievert: Unit and Variety) (1990). He is presently writing his Ph.D. thesis, a study of the Swedish Board of Technical Research and Development.

Index